THE WORLD IS BLUE

HOW OUR FATE AND THE OCEAN'S ARE ONE

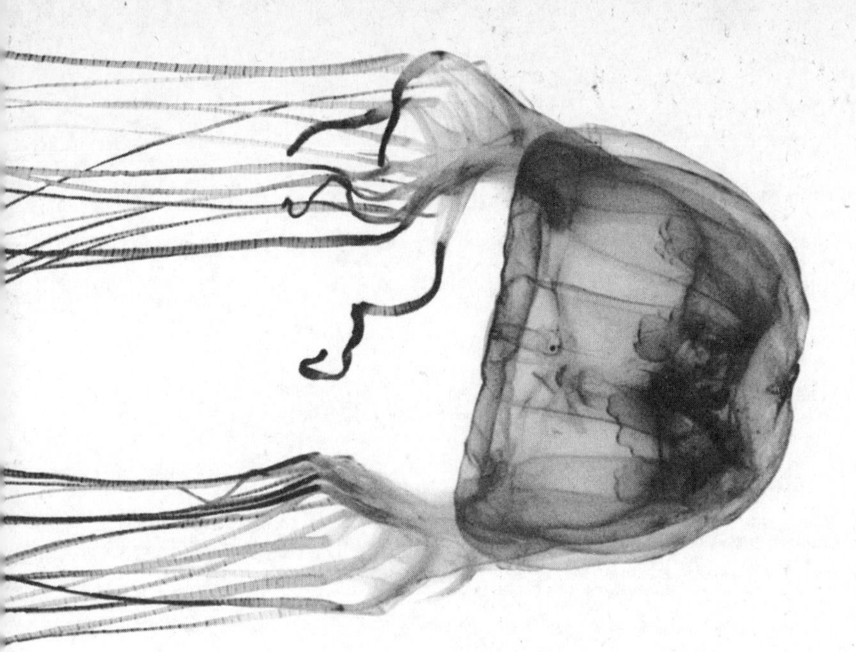

THE WORLD IS BLUE

HOW OUR FATE AND THE OCEAN'S ARE ONE

SYLVIA A. EARLE

NATIONAL GEOGRAPHIC

WASHINGTON, D.C.

Published by the National Geographic Society
1145 17th Street N.W., Washington, D.C. 20036

ISBN: 978-1-4262-0541-5

Library of Congress Cataloging-in-Publication Data
Earle, Sylvia A., 1935-
The world is blue : how our fate and the ocean's are one / Sylvia Earle.
p. cm.
ISBN 978-1-4262-0541-5 (hardcover) -- ISBN 978-1-4262-0558-3 (ebook)
1. Oceanography. 2. Marine ecology. 3. Marine biology. 4. Marine pollution. 5. Nature--Effect of human beings on. I. Title.
GC21.E28 2009
551.46--dc22
2009023972

Photo Credits: 18-19, Gale Mead; 20, Gale Mead; 27, Gale Mead; 32, Gale Mead; 38, Sylvia Earle; 44, Gale Mead; 53, Sylvia Earle; 60, Sylvia Earle; 67, Sylvia Earle; 70, Sylvia Earle; 79, Sylvia Earle; 89, Sylvia Earle; 92, Sylvia Earle; 96, Sylvia Earle; 101, Sylvia Earle; 105, Sylvia Earle; 108, Gale Mead; 109, Gale Mead; 114, Sylvia Earle; 121, Sylvia Earle; 125, Sylvia Earle; 130, Sylvia Earle; 137, Sylvia Earle; 140, Sylvia Earle; 150, Gale Mead; 156, Gale Mead; 165, Gale Mead; 168, Gale Mead; 174-175, Gale Mead; 176, Sylvia Earle; 183, Courtesy NOAA; 190, Tim Taylor; 194, Sylvia Earle; 198, Gale Mead; 209, Gale Mead; 212, Gale Mead; 217, David Guggenheim; 220, David Guggenheim; 227, David Guggenheim; 232, Gale Mead; 237, Sylvia Earle; 246, Gale Mead; 255, Kip Evans.

Maps Credits: Globe: page 8, Carl Mehler, Michael McNey

Graph: pages 280-281
Louisa J. Wood, Lucy Fish, Josh Laughren, and Daniel Pauly. (2008) Assessing Progress Towards Marine Protection Targets: Shortfalls in Information and Action. Oryx, 42(3): 340-351

World Maps: pages 282-285
International Union For Conservation of Nature (IUCN); IUCN World Commission on Protected Areas (WCPA); Sea Around Us Project, Fisheries Centre, University of British Columbia; United Nations Environment Programme (UNEP); UNEP World Conservation Monitoring Centre (WCMC); World Wildlife Fund (WWF)

The National Geographic Society is one of the world's largest nonprofit scientific and educational organizations. Founded in 1888 to "increase and diffuse geographic knowledge," the Society works to inspire people to care about the planet. It reaches more than 325 million people worldwide each month through its official journal, *National Geographic,* and other magazines; National Geographic Channel; television documentaries; music; radio; films; books; DVDs; maps; exhibitions; school publishing programs; interactive media; and merchandise. National Geographic has funded more than 9,000 scientific research, conservation and exploration projects and supports an education program combating geographic illiteracy.

For more information, please call 1-800-NGS LINE (647-5463) or write to the following address:

National Geographic Society
1145 17th Street N.W.
Washington, D.C. 20036-4688 U.S.A.

Visit us online at www.nationalgeographic.com

For information about special discounts for bulk purchases, please contact
National Geographic Books Special Sales: ngspecsales@ngs.org

For rights or permissions inquiries, please contact National Geographic Books Subsidiary Rights:
ngbookrights@ngs.org

Printed in the United States of America.

Interior design: Cameron Zotter

09/RRDC/1

CONTENTS

FOREWORD

If there is one thing most of us know about the ocean, it's that it's big. We may not think about it as often as we should, preoccupied with events here on land, but when we do—well, words like "vast" and "mighty" and "horizon" quickly come into play.

Sylvia Earle, who may know the ocean as well as any human being now alive, helps us cut that intimidating vastness down to size. She does it in two ways, both amply illustrated in this new classic book. Her first method is to bring the ocean to full life—to remind us of the very nearly infinite abundance of things that live there, some of them things that only a few people besides her have ever laid eyes on. Many terrestrial species are still undiscovered, of course, but we're pretty clear on the basic categories: beetles, things with feathers, lizards. Underwater, there are great varieties of stuff we've never even *thought* of, like fish that dangle lights on stalks in front of them to attract their prey. Stuff so big it's hard for us to even quite imagine, such as whales, at once so other and so close. Earle, though a great scientist, is also the heir to Jacques Cousteau, inducting the landbound among us into the mysteries of the sea, helping us to feel both astonished and at ease.

But there's another, much darker, way in which Sylvia Earle helps us understand the size of the ocean. And that's to point out that, vast as it is, it's not so big that we can't screw it up. There's the dead zone at the mouth of almost every big river, and the great gyres of plastic turning slowly in every ocean, and the astonishing fact that we've managed to peer through the opaque depths of the sea with technology that has let us catch most of the biggest fish. We've also managed to melt vast stretches of the Arctic sea, and the UV pouring through the hole we've opened in the ozone has thinned the ranks of Antarctic krill. There's even—and no one predicted this even a decade ago—the fact that we've managed to change the pH of the seas, pouring so much carbon into the atmosphere that they are fast turning acidic and corrosive.

So our job becomes protecting that glory—using the insights and emotions, the reason and the passion, that this tour creates to make us take some serious action. This action is personal—thinking before we order bluefin tuna off the sushi menu. But it is also political, because one by one we're simply not going to change the pH of the ocean back to where it needs to be.

Sylvia Earle's passionate life—including this powerful volume—calls us to that work. But we've got to respond. Brilliant and committed as she is, Sylvia Earle is not going to save the oceans on her own. They're too big. But all of us dwell near the sea, even if we live a thousand miles inland—the sea falls from the sky when it rains; every drop of water we use eventually finds its way into the ocean. It is therefore our duty, and also our delight, to take on this defining challenge of our time.

— Bill McKibben, American environmentalist and bestselling author of *The End of Nature* and *Deep Economy*

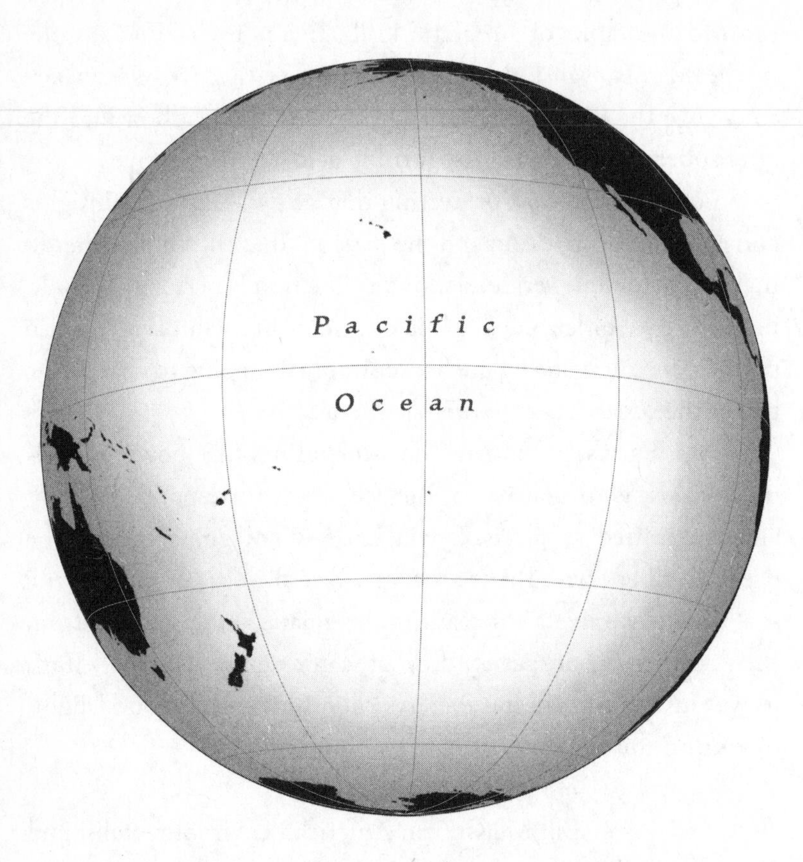

This globe strikingly represents just how "blue" our world is: The Pacific alone covers 58,925,815 square miles of Earth's surface, and that's not even half the total surface area covered by all the world's oceans. Home to millions of species and responsible for driving climate, regulating temperatures, and governing planetary chemistry, the ocean makes life on Earth possible, and its health and protection are issues of critical importance to all of us.

INTRODUCTION: WHY CARE THAT THE WORLD IS BLUE?

Thousands have lived without love—not one without water.

—W. H. Auden

Green" issues make headlines these days, but many seem unaware that without the "blue" there could be no green, no life on Earth and therefore none of the other things that humans value. Water—the blue—is the key to life. With it, anything is possible; without it, life does not exist. Those seeking life elsewhere in the universe focus first on this: *Find the water.* Water is relatively common in the solar system and beyond. Some is ejected from volcanoes through the dissolution of rocks and the fiery union of hydrogen and oxygen, emitted as steam. Comets are mostly rocks and water, essentially dirty snowballs. Jupiter's icy moon, Europa, appears to have an immense amount of frozen water, maybe even some in liquid form under several kilometers of ice.

The presence of water on Mars has been confirmed recently, and it appears that the red planet may once have been blue, with an ocean, now long gone. Only here in this part of the universe, on Earth, is there known to be a place naturally blessed with abundant, liquid water. Not only is this the singular place with an ocean of salt water, but even more significant, it is an ocean that is filled with life that in turn, during some four billion years, has shaped the basic rocks and water of the planet into a strikingly different kind of place, a place unlike any known to exist anywhere else.

Astronomer Carl Sagan noted that even when it is viewed from so far away that it is a pale dot, Earth is discernibly blue. Up close, the ethereal blueness translates to a strange, oxygen-rich, water-laced atmosphere. Densest close to Earth's surface, it gradually thins until merging into space at about 1,000 kilometers (620 miles) above. Below, it sharply converges with the ocean, the salty medium that engulfs all landmasses, large and small. Seen from an orbiting space shuttle, Earth's continents seem to be islands floating in a shimmering indigo embrace, a mass of water now known to occupy 331,441 square kilometers (127,970 square miles) of Earth's surface with an average depth of more than 4 kilometers (2.5 miles), the maximum, 11 kilometers (7 miles) down.

"If the ocean dried up tomorrow, why should I care?" The question, posed by a cheeky Australian reporter in 1976, made me face up to that remote but painful possibility. *Imagine Earth without an ocean.* Gone would be the planet's shroud of salty water, and in reality, even the fresh water in polar ice, lakes, streams, groundwater, and clouds—all soon would disappear without the replenishing mother-source of water, the ocean. Gone, too, would be all of life on Earth. Life can exist in the absence of a lot of things, but as astrophysicist Christopher McKay puts it: "The single non-negotiable thing life requires is water."

But for life in the sea, especially the photosynthetic microbes, Earth's atmosphere would most likely be comparable to that of Mars—more than 95 percent carbon dioxide. Over the past three and a half billion years or so, long before there were mosses, ferns, trees, and flowers, microscopic organisms were churning out oxygen, yielding the atmosphere we now take for granted. The process continues. Without the legions of minuscule organisms that have preceded us over the ages and whose descendants surround us and support us still, life as we know it could not exist.

That's not the only reason we should care that the world is blue.

The ocean drives climate and weather, regulates temperature, absorbs much of the carbon dioxide from the atmosphere, holds 97 percent of Earth's water, and embraces 97 percent of the biosphere. Far and away the greatest abundance and diversity of life occurs in the ocean, occupying liquid space from the sunlit surface to the greatest depths. Even beneath the bottom of the sea, watery cracks several kilometers deep harbor hordes of hardy microbes, which thrive through chemosynthesis, a process of winning chemical energy from surrounding minerals in the absence of sunlight. The sea governs planetary chemistry, yielding water back to the atmosphere that returns to land and sea as rain, sleet, and snow, continuously restoring rivers, lakes, and groundwater.

Even if you never have the chance to see or touch the ocean, the ocean touches you with every breath you take, every drop of water you drink, every bite you consume. Everyone, everywhere is inextricably connected to and utterly dependent upon the existence of the sea.

Early in the 1990s, I listened to Joseph Allen, mission specialist on space shuttle flights STS-5 and STS51-A, talk about how astronauts in training have hammered into them the need to learn everything they can about their life-support system, then do everything possible to take care of it. On the stage was a large photograph of Earth taken during the Apollo program years before. He pointed at it and, with a meaningful grin, said, "Life-support system. There it is. Learn everything you can about it and do everything you can to take care of it."

Earth's life-support system—the ocean—is failing. But who is paying attention? Throughout our history, the mostly blue natural world has been regarded as something to be vanquished, tamed, or otherwise used for purposes that seemed to make sense at the time. Deeply rooted in human culture is the attitude that the ocean is so vast, so resilient, it shouldn't matter how much we take out of—or put into—it. But two things changed in the 20th century that may jolt us into a new way of thinking.

First, more was discovered about the nature of the ocean and its relevance to the way the world works than during all preceding history. Second, during the same narrow slice of time, human actions caused more destruction to ocean systems than during all preceding history. And the pace is picking up.

With hindsight, it is now clear that well before the start of the 20th century, humans had drastically, if unwittingly, altered the fundamental nature of the sea by decimating the populations of fish, mammals, birds, turtles, lobsters, oysters, and other ocean wildlife.

Further changes were initiated by noxious substances lofted into the atmosphere that eventually made their way into the sea. Deliberate dumping of wastes and runoff of agricultural wastes, excess fertilizers, and pesticides already were having a serious impact on the nature of the ocean even before Jacques Cousteau published his pioneering saga about diving, *The Silent World,* in 1953. But since then, the decline has accelerated dramatically coincident with impacts from more than twice as many people as there were half a century ago and development of new technologies that have made exploitation possible in regions previously unaccessible.

Consider:

- Since the middle of the 20th century, hundreds of millions of tons of ocean wildlife have been removed from the sea, while hundreds of millions of tons of wastes have been poured into it.
- Ninety percent of many once common fish have been extracted since the 1950s; 95 percent of some species, including bluefin tuna, Atlantic cod, American eel, and certain sharks have been killed. And taking them is still allowed.

- Destructive fishing techniques—trawls, longlines, rock-hopping dredges—not only continue to take too much, they have destroyed habitats and killed millions of tons of animals that are simply discarded. Every year industrial fishing wantonly kills hundreds of thousands of marine mammals, seabirds, and sea turtles; hundreds of millions of fish, and invertebrate animals.
- Half of the shallow coral reefs globally are gone or are in a state of serious decline since the 1950s; in much of the Caribbean, 80 percent are dead.
- Deep coral reefs are being destroyed by new deep trawling technologies aimed at capturing fish that are decades, even centuries, old. The destroyed corals are thousands of years old.
- More than 400 "dead zones" have formed in coastal areas in recent decades, and the number is increasing and accelerating, reflecting changes in ocean chemistry.
- Global warming and other changes in climate are affecting ocean systems and ocean life just as on land, and those impacts in turn are influencing the atmosphere and terrestrial systems. As the principal driver of planetary climate and weather, changes in the ocean resonate globally.
- The ocean's pH—the measure of alkalinity or acidity—is changing owing to increased CO_2 that in turn becomes carbonic acid. Consequences are likely to be most obvious for coral reefs, mollusks, and plankton housed in carbonate shells, but the changes touch all forms of life in the sea.
- Most troubling, perhaps, is the profound, widespread ignorance about the ocean and its vital importance to everyone, everywhere, all the time. It is not just the fact that less than 5 percent of the ocean has been seen, let

alone explored. Even what is known to scientists is not widely appreciated by the public, and certainly not by most policymaking officials.

So why does the ocean matter?

Of course the economic uses of the ocean matter—extraction of oil, gas, minerals, fresh water and wildlife, transportation, tourism, real estate enhancement, and much more.

Health issues matter. Waterborne diseases are increasing, from cholera linked to coastal contamination and proliferation of toxic algal blooms in coastal waters to mercury and other pollutants in ocean wildlife consumed by the public.

Security issues matter, with competition increasing for rights to use the ocean for military reasons, for access to ports, for fishing, for safe transport of people and goods.

But most of all, it matters that the world is blue because our lives depend on the living ocean—not just the rocks and water, but stable, resilient, diverse living systems that hold the world on a steady course favorable to humankind.

The big question is, what can we do to take care of the blue world that takes care of us?

PERSPECTIVE

For a moment of night we have a glimpse of ourselves and of our world islanded in a stream of stars—pilgrims of mortality, voyaging between horizons across eternal seas of space and time.

—Henry Beston, *The Outermost House,* 1928

I could hear it coming, a sizzling, bubbling rush of green water more than twice my height, a rogue wave that swept me tumbling into a froth of sand and salt water. All I could think of was air! I couldn't breathe, couldn't stand as the undertow pulled me down. I felt deep, numbing fear, when suddenly, the water flowed away, my toes found firm ground, and I stood, sputtering but exhilarated. My mother, up to her knees in surf, started to pull me to the shore, but seeing the look in my eyes, paused and—mother of all mothers—did not stop me when I leaped into the next wave. On a New Jersey beach in 1938, the ocean won my heart. I was three years old.

A wave got my attention, but what held it then and set me on a lifetime course connected to the sea and the creatures that live there were the thousands of craggy, brown horseshoe crabs that clambered ashore every summer. As big as a dishpan, glossy brown with a rounded body, lots of legs, and a long, spiky tail, a horseshoe crab was unlike any other land creature I had ever known. I worried that the crabs would die as I watched them crawling high on the beach, away from the ocean. I spent hours picking them up and putting them back into the water—not knowing that they wanted to go ashore to lay and fertilize eggs in the wet sand, *then* return to the depths. It is a procedure that has worked for these animals for several hundred million years; but, I now know, the horseshoe crab and thousands of other ancient, resilient creatures may not survive the impact my species has had on the living world, largely in a single century. More worrisome, humankind may not survive for long, either, unless we use our remarkable capacity to learn from the past, anticipate the consequences, and take actions that will ensure an enduring future. As it turns out, the future of the ocean, the creatures who live there, and our own future are inextricably linked.

Astronauts return from space transformed by seeing Earth, glowing blue against the infinity of the universe. Similarly, I have been transformed through years of seeing the world from the inside out, first by diving in the rivers and inshore waters of the Gulf of Mexico; later, by leading global expeditions and experiencing thousands of dives into previously unexplored waters. When I first dived in 1952, I was blissfully unaware that in the same year the very last Caribbean monk seal was sighted, an animal once common in the Gulf of Mexico, Caribbean, and Bahamas. I certainly was unaware that the ocean globally was on the verge of cataclysmic decline, that the pristine seas I had known as a child were in danger of becoming Paradise Lost. I was not alone in not knowing. Rachel Carson, famous for her 1962 classic *Silent Spring,* 11 years earlier wrote in *The Sea Around Us:* "Eventually man . . . found his way back to the sea. . . . And yet he has returned to his mother sea only on her own terms. He cannot control or change the ocean as, in his brief tenancy of earth, he has subdued and plundered the continents."

The idea that the sea would hold steady, no matter what we took out of it—or put into it—dominated attitudes and policies globally in the middle of the 20th century. There was already evidence that there were limits to what could be extracted, based on the near extinction of many of the great whales and other marine mammals, turtles, and birds and the serious decline of cod, herring, anchovies, oysters, clams and many other forms of ocean wildlife, brought about by aggressive fishing and the use of massively destructive gear. But the vision of a limitless ocean mesmerized policymakers, encouraging practices that have accelerated the depletion of marine wildlife and minerals; destroyed irreplaceable ocean species and ecosystems; and simultaneously caused the ocean to be regarded as the ultimate Dumpster.

Now, half a century later, we know how wrong we were.

Ninety percent of many fish common when I was a child are now gone, consumed by eager diners unaware that in their lifetime they might witness the disappearance of some of their favorite wild-caught fare, from tuna and swordfish to lobsters and crabs. Low-oxygen dead zones have formed in many coastal areas, largely as a consequence of excess fertilizers and toxic chemicals flowing from upstream fields, farms, and backyards. Plastic debris now clogs beaches, reefs, and even the open sea.

Now we know.

Most worrisome of all is the double whammy of excess carbon dioxide from human activities as the principal driver underlying accelerated global warming and climate change, coupled with the transformation of excess carbon dioxide in the ocean to carbonic acid, now causing acidification of the sea on a grand scale. Perversely, the natural living systems that over billions of years have generated and shaped planetary chemistry in ways that make Earth hospitable for humankind are being destroyed at breathtaking speed.

This, too, we now know.

Now that *I* know, I hope to convey a sense of urgency to others, to inspire use of the special powers that humans possess to take actions to protect what remains and restore whatever we can of the natural living systems that give *us* life, and provide the underpinnings of all we hold near and dear.

I. THE VISION

LIMITLESS OCEAN BOUNTY, INFINITE RESILIENCY

Remains of a whale taken from Antarctic waters.
PREVIOUS PAGES: *A whirlpool of jacks circle the light in Belize.*

TAKING WILDLIFE I—THE MAMMALS

1

Owing to the almost omniscient look-outs at the mast-heads of the whale-ships, now penetrating even through Behring's straits, and into the remotest secret drawers and lockers of the world; and the thousand harpoons and lances darted along all continental coasts; the moot point is, whether Leviathan can long endure so wide a chase, and so remorseless a havoc; whether he must not at last be exterminated from the waters.

—Herman Melville, *Moby-Dick,* 1851

I try to imagine the world as it was in 1900, the year my father was born.

Physically, intellectually, socially, people were much the same as now, but there were no gasoline-powered bulldozers, tractors, or chain saws to help tame the natural systems that preceded cities, farms, and highways. No one had reached the North or South Pole, no canal connected the Atlantic and Pacific Oceans, the first radio broadcast had not yet been received, and few homes were powered by electricity, let alone illuminated by the flicker of television and desktop computers. The discovery of oil in Beaumont, Texas, that triggered an energy revolution was a year away. Railroads stretched across the country, but there were relatively few paved roads and motorized vehicles. Not a shred of plastic anything existed anywhere. A few may have dreamed of people landing on the moon, but the Wright brothers had yet to make a successful flight. Two world wars were yet to be fought, and harnessing nuclear energy for war or peace had not yet become a possibility.

Far fewer people occupied the Earth—less than a quarter of the present population. About 76 million lived in the United States, while China topped all others nations with more than 400 million people (now an amazing 1.4 *billion*). The world's most populous city was London, with 6.5 million residents.

At the time, many believed that the ocean could perpetually yield limitless quantities of fish, whales, and other wildlife, and some still do, although in hindsight, there was and is plenty of evidence to the contrary. The nearshore whales, walruses, seals, and many kinds of fish and tasty shelled animals that had greeted European colonists when they first came to North America had already been depleted, and some, such as the Atlantic gray whale, were gone forever. Compared with the abundance and diversity of life in coastal waters in the 1600s, only fragments remained. Today, there are fragments of those fragments, although to some, what we see today is regarded as normal, as if the present represents the true image of the way things have always been.

EATING OUR WAY DOWN THE FOOD CHAIN

Actually, for 10,000 years preceding the 20th century, humankind had gradually whittled down the natural world, along the way making a living in a manner familiar to most animals—by consuming plant and animal neighbors. Hunting and gathering are activities deeply rooted in human nature and in human culture, underpinning survival throughout most of history. To succeed as a predator, simple math explains that it is vital that the consumers do not outnumber the consumees. The older and larger the consumer, the greater the investment of energy, pound for pound. It takes a lot of seeds and grass to make enough mice and rabbits to make a wolf; a lot of little plants to make sufficient numbers of small fish to make a shark. As it turns out, it takes a lot of everything to power human societies.

In *The Future of Life,* Harvard biologist E. O. Wilson notes how our success has come at the expense of the rest of the living world:

> Humanity . . . eats its way down the food chain. First to go among animal species are the big, the slow, and the tasty. As a rule around the world, wherever people entered a virgin environment, most of the megafauna soon vanished. Also doomed were a substantial fraction of the most easily captured ground birds and tortoises. Smaller and swifter species were able to hang on in diminished numbers.

Never before *Homo sapiens* has a species so comprehensively engulfed the rest of the living and physical world for food, water, minerals, and materials to build and operate the enormous infrastructure that supports civilization. In *Collapse,* Jared Diamond documents patterns of relative success and failure of human societies, depending on their sensitivity to the circumstances that keep them alive. Within a few hundred years of people arriving at the remote and rocky Easter Island, every last tree and nesting seabird was eliminated by a society that essentially consumed itself. For a community or country or civilization to endure, it helps to begin with a robust environment, but what matters most is the approach to living—most significantly, to sustaining the natural systems that underpin all life.

Diamond observes: "Regardless of the resources on which the economy rests—farmed soil, grazed or browsed vegetation, a fishery, hunted game, or gathered plants or small animals—some societies evolve practices to avoid overexploitation, and other societies fail at that challenge." Whether on an Easter Island scale or globally, it is a principle that makes sense.

Long before 1900, it had become obvious that there simply were not enough wild plants and animals on the land worldwide to sustain even a billion people. By then, most of the calories feeding most of the people in the world had shifted from wild things, free for the taking, to a small number of domesticated, cultivated grasses (corn, wheat, rice) and a few kinds of domesticated animals (sheep, goats, cows, pigs, chickens, ducks, and in China, several kinds of freshwater carp). About 100 of the 250,000 or so kinds of flowering plants were being farmed for food, as well as about 20 other bird, mammal, and fish species.

Reliance on farmed food did not stop people from killing wildlife, however. There were still market hunters of ducks, geese, and small mammals, and some communities then relied largely on wildlife—"bush meat"—for much of their sustenance, as they do even now. So many "free" birds and other terrestrial wildlife were still being taken commercially for food, fur, and feathers and for that peculiarly human enterprise, "killing for sport," that laws protecting wild animals began to evolve. Actions came too late to benefit many including the passenger pigeon, a species that a century earlier was said to be the most common bird on the planet, numbering about four billion. The last wild one was shot in 1900. The following year, a *National Geographic* magazine story about Africa lamented, "The fact that the wild animals of the world are in danger of extermination is being forcibly driven home to the minds of all who are interested in natural history . . . This condition is the result of the ruthless persistence with which game of every kind is hunted . . ."

In the same year, the United States passed its first piece of endangered-species legislation, the Lacey Act, making it illegal to move unlawfully killed (i.e., endangered) wildlife across state lines, although the rules did not apply to wild animals with gills and fins. On other fronts, actions were under way in the United States, Canada, Australia, New Zealand, Japan, and elsewhere to establish parks, specific areas aimed at protecting the diminishing natural, cultural, and historic heritage.

HUNTING THE OCEAN'S MAMMALS

The interest in protecting wild places and wildlife was not entirely altruistic. The economic, health, and security benefits of intact ecosystems and relatively stable populations of wildlife were already becoming clear. Curiously, while awareness of the importance of protecting wildlife, wild systems, and places of significance to human heritage was growing with respect to the land, an entirely different attitude prevailed about the ocean. The view expressed early in the 1800s by Lord Byron persisted well into the 20th century:

Roll on, thou deep and dark blue Ocean—roll!
Ten thousand fleets over thee in vain;
Man marks the earth with ruin—his control
Stops with the shore.

Odd that Byron or anyone two centuries ago should think that we lacked the capacity to impact the ocean, or that anything extracted from natural systems comes without cost. But the idea of the sea as a vast, ever resilient source of free food and goods was solidly ingrained, despite clear evidence that the ocean's "big, slow, and tasty" animals—whales, walruses, seals, seabirds, sea otters, cod—and even some small fish species were already in sharp decline. Cod, the fish that sustained North Atlantic nations long before A.D. 1000, by 1800 had been depleted in European waters, while relatively unexploited regions of the western Atlantic were attracting fishermen from thousands of miles away. Disputes over who had the right to capture fish—and where they could be taken—were common, even then.

The detached attitude and lack of concern many have about the past, present, and future of fish, oysters, lobsters, and other sea creatures will be considered later. But what about our fellow mammals? Cattle, sheep, pigs, and other animals farmed for food are all plant-eaters, and all are hustled to market within a year or two of their birth. The older the animal, the greater the investment of room and board—typically about 20 pounds of plants for a pound of farmed yearling cow.

Manatees and dugongs, aquatic relatives of elephants, are plant-eaters too but are slow to mature and have a life expectancy of 50 years or more. The investment of vegetation in making a pound of one of them would fill a warehouse. Other marine mammals—seals, sea lions, otters, dolphins, and walruses—may live for four or five decades; whales, a century (some more than two centuries). All are carnivores requiring for each pound of them thousands of pounds of photosynthetic organisms at the far end of a complex eat-and-be-eaten chain that begins mostly with minute, fast-growing plankton. Those concerned about grabbing and storing carbon from the atmosphere should take note: Every organism, from microscopic plankton to 50-pound tuna or 50-ton whale is a living carbon-based unit, maintaining an enormous "standing crop" of carbon under the surface of the sea. Over time, carbon accumulates in massive deep-sea sediments consisting of the shells, scales, bodies, and bones of trillions of small, medium, and sometimes very large creatures.

Farmers learned long ago that cultivating large, old carnivores for food makes no sense, economically. Taking large, old carnivores from the sea for food makes no economic sense either, if you put ecosystem costs on the balance sheet.

Galápagos sea lion observing the observer.

What happened to marine mammals before 1900 and is happening even now illustrates the illusion of the ocean's limitless abundance, infinite resiliency, and the inclination to keep taking from it long after disastrous declines in original numbers have occurred. Also reflected are the deep roots of humans as hunters, even when killing wildlife for food or commodities is no longer necessary, and even after the critically important value of intact, living species and systems has become obvious.

By 1900, all marine mammals worldwide had suffered serious declines. Seals, sea lions, walruses, manatees, dugongs, sea otters, and polar bears were targeted everywhere for food, fur, and other commodities. A small population of the once wide-ranging Steller's sea cow was discovered in 1741 in the Arctic Commander

Islands by the German naturalist Georg Wilhelm Steller; already severely depleted by aboriginal hunting, the remaining few were gone within 30 years. All other marine mammals present in the 1700s survived to see the 20th century, but several came perilously close to disappearing. The northern elephant seal, sought after for the oil retained in its rolls of wonderfully thick blubber, was thought to be extinct by the 1870s. Several small populations were discovered in Baja California and on Guadalupe Island during the next 20 years, but all that were encountered were immediately killed either for oil or as prized museum specimens. A few, possibly as many as a hundred, eluded the hunters on Guadalupe Island, a haven that became officially protected by the Mexican government in 1922. More than 100,000 descendants of those fortunate survivors have reclaimed a place in Pacific ecosystems.

Sea otters have prospered in kelp forests along the coast of western North America for at least five million years, and they continued to be abundant from Alaska to Baja California even during thousands of years of interaction with native people who hunted them for fur and food. Then, demand starting in the 1700s by European and Russian traders for the otters' luxurious, soft fur pushed them to near extinction within 200 years. In 1910, an international fur treaty finally provided full protection, and small populations at Big Sur, California, and parts of the Aleutian Islands made a tenuous recovery.

Another success occurred in the Southern Hemisphere among the fabled "Robinson Crusoe" islands, Juan Fernández and San Félix, about 640 kilometers (400 miles) offshore from the coast of Chile. As a junior member of a scientific team diving and exploring the region in November 1965 from the U.S. research vessel *Anton Bruun,* I spotted something dark and round bobbing in the water. Someone handed me a dip net, and I scooped up what turned out to be the corpse of a newborn seal pup, umbilical cord still attached. At the time, none of us knew that any of the Juan Fernández fur seals, once numbering in the millions, had survived the comprehensive seal-killing expeditions of the previous two centuries. That year, a colony of about a hundred individuals was discovered not far from where we found the drifting baby; and with full protection awarded in 1978, the population has prospered, now numbering more than 12,000 individuals. While a long way from the millions that once shaped the character of the system with their presence, the trend is promising.

MYSTERIOUS GIANTS

Among all the creatures on Earth, most mammals, especially marine mammals, are giants, compared with, say, insects, shrimp, sea stars, sponges, bryozoans, arrowworms, polychaete worms, and about 95 percent of all other animal life. Following E. O. Wilson's logic that the large forms of life are the first to be eliminated by hungry humans, it follows that the biggest of all—the whales—would be among the most vulnerable. It is likely that the only reason any of the great whales persisted into the 20th century is because of hunters' difficulties getting to where they live.

As a child, I learned to associate whales with whaling, an enterprise that seemed somehow romantic, even heroic. Never having seen a whale myself, it was hard to think of them as creatures with hearts and minds, families and experiences in a realm unimaginably remote from anything familiar to me—except breathing. I did wonder how warm-blooded mammals could live in polar seas, stay submerged so long on a single breath of air, and travel long distances without a map. It was hard to visualize how they could give birth in the sea, how the babies could nurse underwater! Just staying together and not losing one another in the vastness of the ocean seemed somehow incomprehensible. I had come to regard the cats, dogs, horses, squirrels, and rabbits I knew personally as individuals, but I did not think of whales in the same way. And at the time it did not occur to me to question the reasons why people seemed intent on killing whales.

The turning point for me came from reading passages for a class in biology in the 1874 book *The Marine Mammals of the North-western Coast of North America* by Charles Scammon, a whaling boat captain who not only led expeditions to kill whales, but as a keen observer and naturalist, sought to record information about them. In Scammon's time, no one intentionally got into the water with whales, so it was difficult to even know what they looked like in their own realm.

Scammon never came close to capturing the full and accurate information. He notes, "close observation for months, and even years, may be required before a single new fact in regard to their habits can be obtained . . . it is extremely difficult to delineate accurately the forms of the larger Cetaceans. . . . but a small part of its colossal form can be seen, as, usually, only a small portion of the middle section of the body is above the water." He saw plenty of dead whales, but one sprawled misshapen on the deck of a ship is as much like a whale in its element as a wood pile is to a living tree in a forest.

In Baja California, Scammon was surprised to see a number of gray whales that seemed to be bodysurfing. "One in particular," he wrote, "lay for a half-hour in the breakers, playing, as seals often do in a heavy surf; turning from side to side with half-extended fins . . . at times making a playful spring with its bending flukes, throwing its body clear of the water, coming down with a heavy splash . . . to all appearances enjoying the sport intensely."

So, whales play? Have individual personality? When I read this, I suddenly saw whales in a new light, as amazing multifaceted beings, and whalers were no longer heroes. I immediately went about learning more about these magnificent creatures.

Worldwide, there are about 80 kinds of whales and their smaller cousins, dolphins and porpoises. A dozen or so, the Mysticeti, include the largest whales—blue, fin, Bryde's, sei, humpback, and gray as well as the petite minke, about the size of a large dolphin. The rest, the Odontoceti or toothed whales, range from the 60-foot-long sperm whale to New Zealand's hold-in-your-arms–size Hector's dolphin and Mexico's diminutive vaquita. Most are entirely marine, but a few became adapted to living in rivers in India, China, and Brazil.

Knowing about whales is one thing; knowing how to kill them is another, an interest that reflects our deep roots as hunters. Scammon describes weaponry and killing strategies in exquisite detail, including the clever method employed to get close to whales: Harpoon a curious baby to attract the mother, then harpoon her. According to Scammon, the parent sometimes "in her frenzy, will chase the boats, and overtaking them, will overturn them with her head, or dash them to pieces with a stroke of her ponderous flukes." Such behavior earned gray whales the name devilfish. There is no record of how the whales referred to the whalers.

So, whales care for their young and will defend them at their own peril? This was another reminder that whales are not just commodities.

Another shift in my perspective came in 1976 when I listened to biologist and whale expert Roger Payne talk about his experiences with southern right whales in Patagonia. He identified individuals by their distinctive faces: Each bears characteristic patterns of rough skin known as callosities. Not surprisingly, each whale behaves differently, too—a discovery made by Jane Goodall about chimpanzees; by Dian Fossey with mountain gorillas; by essentially everyone who has carefully studied any creature, from human beings to beetles to bears, from eels to elephants. It turns out that no creature, no tree, no branch of any tree, and certainly no whale is exactly like any other. The capacity for such staggering diversity seems to me to be one of the two great

Common dolphins such as this one are common no more.

miracles of life. The other is the common water-based chemistry that unites all.

After his lecture, Roger and I talked about an idea he had for correlating the behavior of humpback whales to the melodious sounds they make, by observing, recording, and filming them—underwater—something no one had tried before. We agreed on the spot to work together to organize an expedition to explore the possibilities; and set about winning support from several sources—the National Geographic Society; the New York Zoological Society (now the Wildlife Conservation Society); the California Academy of Sciences; Hawaii's Lahaina Restoration Foundation; the National Oceanic and Atmospheric Administration; and a British film company, Survival Anglia. A few months later, on February 13, 1979, submerged in the open sea several miles offshore from Lahaina, the old whaling port in Maui, Hawaii, I met a live whale face to face for the first time.

Four of us in a small rubber boat had been admiring rainbows forming against the misty breath of five whales, when they abruptly changed course and swam straight for us. We stopped the motor, and with photographers Al Giddings and Chuck Nicklin I carefully slid into the sea, hoping to catch a glimpse of the big animals as they passed by in the crystalline channel water. I had told a lot of people who were backing the expedition that I wanted to observe whales on their own terms, underwater, but it hadn't occurred to me that the whales would be observing me as well.

Sleek and supple, 40 tons of what appeared to be an enormous torpedo sped toward me on a clear collision course. With no time to climb back in the boat and no place to hide, I froze. Whatever happened next would be up to the whale. With a subtle twist of her fluke, she turned, lifting her winglike flipper to avoid contact, and swept by, her eye moving slightly, acknowledging my presence. She next inspected Giddings, coming so close that he felt the rush of water as she passed. For two and a half hours she and four other whales engaged us, rushing headlong in our direction; disappearing beneath us into deep, ethereal blueness; then rising like synchronized dancers, spiraling to the surface with gargantuan grace. In *Moby-Dick,* Herman Melville describes humpbacks as "the most gamesome and light-hearted of all the whales, making more gay foam and white water generally than any other of them." Watching them underwater, I would add, "and the most gloriously misrepresented."

WHALING NATIONS

Whales portrayed in Captain Scammon's book and others of that era suggest swollen buses or giant loaves of bread, always level, as if fixed in two dimensions, nothing like the lithe gymnasts Melville described or that we witnessed underwater in Hawaii. Years of later observations suggest that we were swept up in the midst of amorous suitors courting a lone female, but at the time, few had dived with humpbacks or any other large whales, so we had no way of surmising this. No one knew the nature of their social interactions, nor could anyone predict how they would react to the presence of people in their midst. Humpbacks may live longer than humans, and the whales we were with may have encountered not-so-benign members of our species wielding harpoons decades or even a century ago. At the time of our dive, Japan, Russia, Iceland, Australia, South Africa, and New Zealand were among the nations still killing whales for meat, oil, and bone. In the United States, humpbacks, grays, Bryde's, fin whales, blue whales, and, occasionally, orcas, were taken commercially until the mid-1960s, but the last whaling station, based in Richmond, California, did not close until 1971.

In the 20th century, about three million whales, including more than 300,000 humpbacks, were killed by whalers from 46 countries. Norway took the most, 27 percent, with Japan taking 21 percent, the U.S.S.R. 18 percent, and the United Kingdom 11 percent. Whether this seems like a lot or a little, the fact is that all of the great whale species were reduced to a small fraction of their pre-whaling numbers. The gray whales described by Scammon in Mexican lagoons came close to complete annihilation.

In the summer of 2008 on an island in the Norwegian archipelago of Svalbard, I walked among the mossy bones of bowhead giants, all that remained of the animals whose bodies yielded what some refer to as the first great oil rush. Pink and white blossoms sprouted from dark green mounds of Arctic wildflowers, darker and greener closest to where ribs, jawbones, and skulls still leaked nutrients derived from long-ago ocean food chains. By the end of 1600, several thousand ships from several European nations worked the waters from where I stood to Greenland, eventually shifting from shore-based operations to at-sea processing as the close-in whales were killed off. By 1800, bowhead and right whales throughout their range were reduced to a fraction of their former numbers, but the killing continued.

Attention turned to blue whales, *Balaenoptera musculus,* the largest animals ever to live, but others—fin, sei, humpback, and gray—were targeted as well. The methods of traveling long distances to find, catch, render, and market whales advanced quickly with the advent of steam- and diesel-powered vessels in the 19th century, and the invention of an exploding harpoon gun in 1868 greatly increased the ease and certainty of killing targeted animals. Nothing in the 65-million-year history of whales had prepared them for the level and mode of predation imposed by a terrestrial species equipped with seagoing machinery, a method that had the potential for eliminating every last individual.

Incredibly, despite the near disappearance of several species, whaling was still big business at the start of the 20th century. By the mid-1900s, only about 10 percent of the blue whales remained of a global population estimated to be 275,000 a century earlier.

In 1946, in response to obvious declines, whaling nations formed the International Whaling Commission (IWC) to regulate the numbers taken—and thus, in theory, to preserve the whales and so the whaling industry. An ingenious method was developed to make up for the loss of blue whales, the creatures that yielded the greatest return per animal in terms of barrels of oil. One blue whale could net about 120 barrels, a fin whale about half that amount. With cool, mathematical logic, a Blue Whale Unit (BWU) was equal to two fin whales, two and a half humpback whales, or six sei whales. In 1966, the IWC banned the killing of blue whales, and some restrictions were placed on other species. But the following year, 67,000 whales of various kinds were taken—more than twice the number reported killed in 1933, when there were no limits on taking.

Especially valuable were the highly social sperm whales, notable for having the biggest brain of any animal, ever. The pale, waxy spermaceti contained in their massive head was prized as a fine lubricant, as was ambergris, a dark, aromatic substance formed in their stomach, a precious ingredient in certain perfumes. Teeth were kept as ornaments and for carvings, but after a layer of fatty blubber was stripped away, the remainder of the body was usually discarded.

In a drizzling rain at dawn in 1978, I witnessed the bodies of four young male sperm whales off-loaded for processing by the Cheynes Beach Whaling Company in Albany, Australia; it was the last year of that nation's whaling operations. A huge power saw decapitated each whale, then a dozen men in sulfur yellow slickers wielding long-handled blades efficiently sliced the sleek, gray forms into pieces small enough to be threaded through six gaping holes in the blood-slick flensing deck and into steaming vats below. Meat and bone were cooked under pressure, ground to a paste, then pumped into settling tanks to separate the oil from all the rest.

The week before, the whales were with others in their age group, diving 1,000 meters (3,300 feet) or more on a single breath, using sonarlike clicks to find squid and other prey, experiencing pressures lethal to humans, and witnessing places humans have never seen—broad areas of dark wilderness illuminated solely by the eerie blue-green flash, sparkle, and glow of thousands of bioluminescent creatures. Biologist Hal Whitehead describes the sounds of sperm whales in *Voyage to the Whales,* his fine account of observing sperm whales in the Indian Ocean from a small sailboat. Through headphones attached to underwater microphones, lowered hundreds of feet below the surface, he hears a rhythmic knocking that pauses, then "restarts at a faster rate and then accelerates, turning into a lingering creak . . . two other sperm whales add their reverberant clicks. . . . Several more whales join in, and now the chorus sounds like a horse race on hard ground." Years before, whale expert William

Watkins discovered that each sperm whale has a distinctive set of clicks that he called codas, referring to a term used to describe a special section of a musical composition. Upon encountering another whale, codas are exchanged, evidently a whale's method of saying "Hi! I'm Sam." No one knows the full extent of how these and other toothed whales use their built-in sonar system, but according to a theory posed by biologist Kenneth Norris of the University of California at Santa Cruz, intense pulses of sound produced by sperm whales may be used to find and stun squid and other prey.

It takes 25 years or so for a male sperm whale to reach the edge of social maturity, when it attains the size and weight of those I saw being butchered. It took less than four hours to transform those once vibrant creatures into the basic ingredients of candle wax, lubricating oils, cosmetics, fertilizer, ivory trinkets, and food for domesticated animals.

Australia now leads antiwhaling efforts globally, but in the mid-1970s, the hundred or so people whose livelihoods were based on killing sperm whales caused the country to be classed as a whaling nation at the annual IWC meetings, where members deliberate and establish quotas on where and how many whales of each type may be taken, and by whom. Since 1978, Australia has ceased killing whales, but whales continue to provide income in Albany and Twofold Bay where museums devoted to whales and whaling are popular tourist attractions.

Although no longer engaged in commercial whaling, the United States is still technically a whaling nation, owing to bowhead and gray whales' being taken for "indigenous cultural purposes" in the waters offshore from Washington and Alaska. As a member of the IWC for four years and deputy commissioner for the United States for two, I participated in many discussions with the representatives from Japan, whose determination to take whales commercially has created potent international political

Walruses are feeling the pressures of hunting and loss of Antarctic ice.

discord that overwhelms rational analysis, global public opposition to whaling, and volumes of scientific data.

In the 1990s, Kuzno Shima, head of the Japanese delegation, was particularly provocative, posing questions to me such as "Americans eat beef, right? What's the difference between eating a steak from a cow and eating whale meat?" I tried to respond seriously: Cows are herbivores and go to market in a year or two, have been cultivated by people for food for ages, and require care and an investment of some sort by farmers; while whales are free, wild beings that belong to no one, are typically taken after they have lived for decades, and are relatively few in numbers (or are not "restocked" like cows), leaving an irreversible tear in the ocean's fabric of life when removed. There are billions of cows, but all whale species are greatly reduced in number, some bordering on extinction owing to whaling. Taking even a few increases the risk of depletion owing to other pressures—storms, disease, pollution, and fluctuating food sources. The whales of today have ancestral roots 65 million years deep, and nothing in their survival strategies factored in the impact of humans as predators. What might we learn from them as living creatures, able to communicate with sound over long distances, develop close-knit societies, navigate over thousands of miles with no map, and perform daily deep-diving feats that defy the capacity of even the most athletic humans? If only considering whales as a priceless source of knowledge, we discover that their value alive far exceeds their worth as pounds of meat. In narrowly defined economic terms, the growing business of whale-watching is lucrative and demonstrably sustainable, while commercial whaling is subsidized, with a consistent record of "management" failure.

Shima listened politely to my naively earnest summary of how whales differ from cows and why live whales are more valuable than dead ones, but he just smiled and said it was traditional to hunt whales and insisted that it could be done sustainably. He justified killing whales for scientific analysis as a necessary way to better understand them. He seemed immune to my question about what aliens would know of human society, poetry, music, language, behavior, or aspirations if they examined only dead bodies. Perhaps Shima was motivated by national pride or concern about precedent-setting policies that could spill over into fishing. Whatever the issue, he clearly shared the view expressed by a man I met in Tsukiji, the renowned fish market in Tokyo. When I asked what he thought about whales, he beamed and said, "Ah! Delicious!"

Residents of the Danish Faroe Islands have similar ideas about the pilot whales that once provided a primary source of sustenance. Though no longer needed for survival, they are still eaten. In a macabre "tradition," hundreds are still herded into shallow embayments for slaughter with clubs, spears, and lances. Men wade waist deep among frantic whales in water that soon flows red with the blood of the dying animals.

DOLPHINS IN DECLINE

Small numbers of dolphins have been taken for food in some countries for centuries, but the massive scale of taking striped, spotted, bottlenose, and Risso's dolphins, as well as pilot and false killer whales in Japan in the 20th century, and even now, is accelerating the decline of these species. As in the Faroes, several shallow bays in Japan, Peru, and the Solomon Islands are used to confine dolphins for similarly brutal modes of killing. A film, *The Cove,* shown at the Sundance festival in 2009, electrified viewers with documentation of how Japanese whalers at Taiji, Japan, frighten wild dolphins inshore with loud noise, cut off their escape with nets, then drive them together for death by knives. Lured by the promise of "something for nothing" from the wild, meat from the animals is sold for consumption in Japan; a few animals are set aside for sale, alive, for exhibition as tourist attractions.

Deliberate killing is not the only kind of grief humans have imposed on dolphins in recent times. From 1950 to the 1990s, more than six million spotted, common, and spinner dolphins died in the eastern tropical Pacific alone as "incidental bycatch" by fishermen using purse seines to capture yellowfin tuna. The dolphins associate closely with the tuna and are swept up with them when the seine is brought to the surface. It is surprising not that dolphin populations dropped sharply in the 20th century, but rather, that there are any left.

What is even more surprising is that despite the war we appear to have waged on dolphins in recent years, they seem to bear humankind no significant ill will. The Greek biographer Plutarch, who lived near the Mediterranean Sea, observed centuries ago: "To the dolphin alone, nature has given that which the best philosophers seek: friendship for no advantage. Though it has no need of help from any man, it is a genial friend to all."

Most of my personal underwater encounters with several species of dolphins in the Atlantic, Pacific, and Indian Oceans have been chance meetings, the dolphins occasionally showing brief curiosity or benign indifference—with a few memorable exceptions. In the northern Gulf of Mexico, while cruising in shallow water in the one-person submersible *Deep Worker,* several bottlenose dolphins, *Tursiops truncatus,* appeared from the surrounding green haze and circled the sub, peering into the clear dome that covered my head. As I continued on my way, they stayed, matching my slow-motion pace for several minutes, then racing ahead, circling around, returning again, then racing away. Like my family's Chesapeake Bay retrievers on a walk, the point-to-point distance we covered was the same, but the dolphins traveled at least five times as far, not counting their excursions to the surface to grab breaths of air.

I was skeptical about reports that a "friendly dolphin" had appeared along San Salvador Island in the Bahamas in 1978, but from time to time, there are stories about wild bottlenose or other dolphin species playfully interacting with people. I was intrigued enough to give my three children, 8-year-old Gale, 14-year-old Richie, and 16-year-old Elizabeth, a break from their schoolbooks to accompany me on an assignment from the National Geographic Society to check out the dolphin story. In a small boat, we cruised along the shoreline where we were told to go, and suddenly, there he was, a wild spotted dolphin, *Stenella frontalis,* locally known as Sandy, arching over low, rippling waves right for us. We anchored, and one by one, slipped into the clear, warm water as the dolphin circled around, eyeing us while emitting staccato beeps and whistles. He seemed to prefer the children, approaching each one as if inviting them to play, and they responded with arching twists and turns, a happy mammalian muddle of bodies, fins, and arms.

WHAT IS TRULY SUSTAINABLE?

The immense goodwill the great majority of people feel toward dolphins, whales, and other marine mammals is offset by the attitude of those who continue to regard them as "pests," i.e., competitors for fish, or as "resources"—commodities to be gathered and traded.

Some claim that it is possible to take such animals sustainably. In this view, that annual replenishment will balance those that are removed, a process that can continue indefinitely as long as sufficient numbers remain to reproduce the next generation at a somewhat stable level. It is a seductive theory, one that inspires ingenious models to demonstrate how exploited populations can recover year after year.

Despite good intentions and sound logic, however, natural populations tend not to follow the rules. Part of the problem is knowing how many of any species exist at any one time, let alone how many existed before serious exploitation got under way. Consistent, regular monitoring is rare and difficult to accomplish, but vital: How else is it possible to tell what's going on? In addition to natural variables, there are other human factors—pollution, habitat disruption, the serious losses incurred by entanglement in lost fishing gear, as well as illegal taking that exceeds official limits. Most models factor in contingencies, but in the wild ocean, it is impossible to know, let alone control, the predictably unpredictable—a topic that will be addressed in the next chapter.

It could be argued that Norway, second only to Luxembourg in per capita income, cannot justify its insistence on whaling on the grounds of sustenance or immediate economic need. The country's former prime minister, Gro Harlem Brundtland, assured me during a meeting in Monaco in 2009 that "Norway's whaling policy is not about the economy; it is about sustainable development."

Interpretations of "sustainable development" vary, but the most widely accepted and applied definition appears in the Report of the Brundtland Commission, *Our Common Future,* published in 1987: "Sustainable development is development that meets the needs of the present without compromising the ability of future generations to meet their own needs."

The concept is brilliant, and it conveys an ethic consistent with the idea that all of us should do whatever we can to leave the world as good as—or better than—we find it, to use natural resources without using them up. "Sustainable development" as applied to wild animals is tricky, however, and with respect to the nations (Japan, Norway, and Iceland) currently taking minke and other whale species on a commercial scale, it suggests an obligation to kill whales as a matter of principle, whether or not they are actually needed for food or income.

Whatever the rationale for killing the 120 or so species of marine mammals, the 20th century was devastating for most species, but there is good news: With one exception, some of every species around in 1900 lived to greet the year 2000. Only the Caribbean monk seal, common when Christopher Columbus came to the Americas late in the 1400s, present in abundance when Ponce de León came to Florida in the 1500s, and common enough through the early part of the 20th century to cause fishermen to kill it as a potential competitor for fish, was last seen in 1952.

Imagine monk seals lounging on beaches in Nassau, Cuba, and Miami! As a child living on Florida's west coast, it never occurred to me that seals might have been my neighbors, splashing around in nearby waters. Cayo Lobos and the Seal Cays in the Caribbean are bleak reminders of what was but can never be again. It may not be too late to save their warm-water cousins, Mediterranean and Hawaiian monk seals, now fully protected. A few hundred of each remain, symbols of hope for the survival of their species—and maybe, for ours.

Fish have faces too: Note this sweetlips on a reef in Palau.

TAKING WILDLIFE II—THE FISH

Who knows what admirable virtue of fishes may be below low-water mark, bearing up against a hard destiny, not admired by that fellow creature who alone can appreciate it! Who hears the fishes when they cry? It will not be forgotten by some memory that we were contemporaries.

—Henry David Thoreau, *A Week on the Concord and Merrimack Rivers,* 1849

Long before the 20th century, taking large quantities of marine mammals had proved to be unsustainable. More surprising, by the time the century began, populations of nearshore fish had also declined significantly. However, many people were convinced that fish could rebound, no matter the number caught. Even the distinguished British scientist Thomas Huxley was skeptical about the ability of people to appreciably diminish the number of any kind of fish, as he famously said in an 1883 speech:

> The herring fishery, the pilchard fishery, the mackerel fishery, and probably all the great sea fisheries are inexhaustible: that is to say that nothing we do seriously affects the

> numbers of fish. Any attempts to regulate these fisheries seems consequently . . . to be useless.

Fishermen loved the idea that the ocean would never run out of fish, but alas, the fish were not paying attention to Professor Huxley. By 1880, Atlantic halibut populations had collapsed, and the species remains rare today. Cod, for several centuries the economic mainstay of several European countries, was depleted in the North Atlantic even before Huxley's time, and herring? Well, to Huxley, "herring" meant what scientists call *Clupea harengus harengus,* the Atlantic herring, one of about 225 small, silvery fish species in the family Clupeidae that occur in massive schools in temperate waters globally. They, too, had fallen in abundance in the North Atlantic by the late 1800s, owing to heavy fishing pressure from European countries as well as from fishermen in Canada and the United States. Yet they were—and still are—numerous enough to inspire continued catching, providing sustenance and products for people, a very recent addition to a long list of traditional consumers—other fish, whales, seals, seabirds, squid—most of which are wholly dependent on the existence of large quantities of small fish. As we take more, they, of course, get less.

THE TRAGEDY OF THE COMMONS

In retrospect, it should have been obvious that there are limits to how much can be extracted without undermining future resources, especially in a realm from which everyone is free to take as much as they want using whatever means they choose. The Tragedy of the Commons, economist Garrett Hardin called it in an often cited paper in *Science* in 1968. As an example, Hardin imagined an open pasture where people could graze as many cows as they wished. There is incentive for each to have as many cows as possible, the rationale being, "If my cows don't eat the free grass, somebody else's cows will." It is to the advantage of each to have more cows eating more grass than the next person. The grass is free! But in time, there are too many cows, too little grass, and no more of anything for anybody. The end point, according to Hardin, is ruin—"the destination toward which all men rush, each pursuing his own best interest in a society that believes in the freedom of the commons."

In the 19th century, there was a rush to catch whales, seals, and seabirds from the global commons (they were free for the taking!), with resulting tragedy: near extermination of many species and the loss of a way of life for whalers, sealers, and bird-hunters. Taking mammals and big birds from the sea continued through the 20th century into the 21st, but the principal rush for wealth from the sea focused on fish and "shellfish"—oysters, clams, lobsters, shrimp, and other invertebrates. Although some species of fish already had suffered sharp declines by 1900, the total amount extracted globally in that year was about three million tons. As technologies improved and global markets expanded along with increasing populations, the rush to acquire from the sea what was "free for the taking" was on, and the "take" increased accordingly. At least for a while.

Populations of commercially caught fish got a break during World War I, but by 1919, new and larger fleets were launched. Increased fishing led to increased tonnage of fish landed, but within a few years catches dropped as populations were depleted. In an effort to figure out what to do to maintain high levels of taking, ideas arose about managing fishing on a scientific basis. Until then, it had seemed reasonable to suppose that when one species declined to the point where it was no longer profitable to fish for it, other species would be targeted while the depleted one recovered in an endless cycle of extracting a "renewable" resource.

Biologist Harden F. Taylor echoed this view in 1951 in the *Survey of Marine Fisheries of North Carolina:* "Heavy fishing may reduce a population, but the fishery arrests itself automatically as it becomes unprofitable and is discontinued or much diminished long before any species is totally 'fished out.' "

THE MYTH OF MAXIMUM SUSTAINABLE YIELD

In the 1930s, a seductive strategy for managing fisheries emerged that aimed to extract the largest possible catch from a stock of fish over an indefinite period of time. Called maximum sustainable yield, or MSY, the idea presumes that a population is able to reproduce at its maximum efficiency when it is reduced to about half the number that can be sustained in a given area. In theory, the remaining individuals will produce a surplus that can be skimmed off year after year, with the population overall remaining steady. Calculations are based on the number presumed to be present before fishing began. It assumes that growth rates, survival rates, and reproductive rates are maintained normally, and in fact, are enhanced when the population density is reduced by fishing. In other words, it is a good thing for the fish to lose half their population. They respond, in theory, by instantly making a lot more fish!

So simple and appealing was the concept that, between 1949 and 1955, maximum sustainable yield became the goal of international fishing management policies. MSY was adopted as a primary management target by the International Whaling Commission, the International Commission for the Conservation of Atlantic Tunas, and other fisheries organizations.

World War II had given ocean wildlife a second wartime break, but the application of MSY principles encouraged fishermen to increase their efforts and take populations down to what were thought to be sustainable levels. The idea persisted that if there were fewer fish, they would have more space and more to eat, and would reproduce faster than before populations were reduced.

But those pesky animals didn't obey the rules. So, what's wrong with the concept of maximum sustainable yield?

First, it would help to ask the fish. At that time, critical aspects of their life were unknown, and in fact are still largely a mystery. It is really tough to predict the future behavior of an animal if you don't know much about its past or present.

Second, it is difficult to determine how many fish of a particular species are in an area at any one time, let alone how many there were before fishing began. Imagine estimating the population of New York City from an airplane flying thousands of feet overhead in a dense fog. You have to lower a net to snag a sample and extrapolate from that sample how many people there are in the entire region. To make the sample size larger, you can do it again and again, thereby destroying buildings, raising havoc in the city, and reducing the size of the population even before serious catching begins. Modern assessment sonars and camera systems have impaired the resolution of where and how many of what kind of fish are in an area, but they have also tipped the scales in favor of the fisherman. As manufacturers of "fish finding" equipment advertise, "the fish have no place to hide."

Third, fish populations undergo natural fluctuations, impossible to predict given the magnitude of what is not known about the life histories of species, the nature of the environment, and how the individual species and its surroundings interact. A sample taken one day in one area may not reflect what is—or is not—present year-round or over a wide area.

Fourth, there is no surplus in a natural, healthy system. What appears to be overabundance to human observers is a natural insurance policy against population reduction by disease, storms, ups and downs of predators and food supply, temperature shifts, and other aspects of life beyond a computer model.

Fifth, species do not live in isolation; They are integrated into exceedingly complex systems composed of thousands of other forms of life, each undergoing constant change. Since less than 5 percent of the ocean has been seen, let alone explored, many of the systems remain obscure. But, as in a human city, it takes everyone doing lots of different things for the system to work for each citizen. Removing just the taxis or trash collectors or knocking out a few bridges could be disastrous.

Sixth, successfully plucking individual species out of a system is difficult. Trawls indiscriminately scrape up all species in their path, as well as pulling up the habitat itself. That, in turn, undermines the capacity of the system to continue producing fish according to plan. Baited hooks attract more than the desired species, causing an "incidental catch" of birds, mammals, turtles, and numerous unintended fish. Drift nets, like trawls, take everything too large to get through the mesh. Traps for fish, lobsters, and crabs capture and kill many other kinds of fish and invertebrates. Imagine trying to selectively extract just the lawyers out of New York City!

Seventh, no individual fish, let alone an entire species, lives, eats, behaves, grows, or reproduces precisely like any other. Models such as MSY are useful to anticipate possibilities but rarely predict reality.

Eighth, the first fish to be taken from an unexploited population are the largest and oldest, the "old timers" that have demonstrated superior survival traits. They are also the ones that produce the most offspring. When they go, hard-won experience as well as reproductive capacity goes, too. In effect, the kids are left in charge.

Ninth, most fish do not grow to maturity within a year. Herring take four years to begin reproducing and may live for 20 years;

bluefin tuna take eight years to mature and may live to be 30; orange roughy mature at about 30 years and may live more than 150 years. Meanwhile, fishing pressure is continuous and relentless.

Tenth, political expediency, not the well-being of fish populations, distorts the application of even well-intended policies.

Eleventh, people want to believe that MSY works, and keep on believing that it works, even when experience demonstrates that it *doesn't* work.

The twelfth, and perhaps most important, flaw in the MSY concept is that it regards fish and other ocean wildlife first and foremost as commodities, with an implied obligation to take them as such. The important functions of intact ocean systems that benefit people everywhere (generating oxygen, taking up carbon, maintaining biodiversity, driving the water cycle, shaping planetary chemistry, holding the planet steady, and so on) are put aside in favor of single-minded extraction of salable products, benefiting relatively few.

Biologist P. A. Larkin described the concept of MSY in 1976 in a keynote address to the American Fisheries Society as an optimistic dogma that assumes that "any species each year produces a harvestable surplus, and if you take that much and no more, you can go on getting it forever and ever." Recognizing the fatal flaws, he was inspired to suggest that MSY should be buried with the following epitaph:

M.S.Y.
1930s–1970s
Here lies the concept, MSY.
It advocated yields too high,
And didn't spell out how to slice the pie.
We bury it with the best of wishes,
Especially on behalf of fishes.

We don't know yet what will take its place,
But we hope it's as good for the human race.

In a serious attempt to come to grips with the problems inherent in MSY, two seasoned ecologists—Sidney Holt, with years of experience in fisheries biology, ecology, and politics with the Food and Agriculture Organization (FAO), and Lee Talbot, a globally knowledgeable wildlife ecologist and conservationist—wrestled with the issues in a thoughtful analysis in 1978. Had their proposed "New Principles for Conservation of Wild Living Resources" been applied, the state of wildlife, land and sea, would be much more robust. The principles still make sense:

> The privilege of utilizing a resource carries with it an obligation to adhere to the following . . .
>
> 1. The eco system should be maintained in a desirable state such that
>   a. Consumptive and non-consumptive values could be maximized on a continuing basis
>   b. Present and future options are ensured
>   c. Risk of irreversible damage or long-term adverse effects as a result of use is minimized
> 2. Management decisions should include a safety factor to allow for the facts that knowledge is limited and institutions are imperfect
> 3. Measures to conserve a wild living resource should be formulated and applied so as to avoid wasteful use of other resources
> 4. Survey or monitoring, analysis, and assessment should precede planned use and accompany actual use of wild

A giant manta ray flies over a reef in Raja Ampat.

> living resources. The results should be made available promptly for critical public review

Unfortunately, the new principles were applauded but not adopted. Far from being abandoned, the MSY doctrine became solidly embedded in policies throughout the world. In 1982, it was incorporated into the United Nations Convention on the Law of the Sea, and it has been integrated into national and international laws ever since, sometimes dressed up in such variations as "optimum sustainable yield" and "long-term potential yield."

Optimism about the ability to extract large quantities of ocean wildlife on a sustainable basis continued to shape fishing policies throughout the 20th century. Despite depletions in some areas and for some species, total catches increased until the 1980s, owing to increased effort and greatly enhanced means of finding and catching previously untapped populations—the remaining "old growth forests" of fish.

FROM SUNLIGHT TO PLANKTON

To understand why it is difficult for us as newcomers to sustain large-scale extraction of ocean wildlife from ancient ecosystems, and to see how commercial fishing is linked to changes in planetary chemistry, it helps to consider the basic processes involved.

The conversion of energy from sunlight to photosynthesizers to animals occurs first on the micro-microscopic scale, and continues thereafter for the lifetime of the consumer, the diet changing with increasing size. Much of the action starts with sunlight, converted in the cells of trillions of microscopic organisms supplied with chlorophyll, carbon dioxide, and water, into simple sugar and oxygen. This is how a large measure of atmospheric carbon dioxide is taken into the sea, and how a large measure of oxygen is discharged back to the atmosphere. One kind of blue-green bacteria, *Prochlorococcus,* is so abundant—about 100 octillion (1 octillion = 10^{27}) are alive at any given moment—that it alone is responsible for about 20 percent of the oxygen in the atmosphere. Put another way, this nearly invisible form of life generates the oxygen in one of every five breaths you take, no matter where on the planet you live.

Other ocean photosynthesizers contribute an additional 50 percent of our atmospheric oxygen. Clearly, we would all have a hard time breathing if we had to rely on trees, grass, and other terrestrial plants alone for oxygen.

Like most other microscopic phytoplankton, *Prochlorococcus* cells grow rapidly, with a lifetime measured in hours or days. Most are quickly consumed by minuscule grazers, from browsing single-celled protozoa to the larval and adult stages of numerous categories of animals, such as crustaceans, sponges, polychaete worms, arrowworms, sea stars, even hatchling herring. Other tiny but nutritious life-forms—planktonic diatoms, dinoflagellates, cyanobacteria, and most other floating photosynthesizers—are much too small to be eaten directly by most large creatures. The very young stages of many species may ingest phytoplankton, but as they grow, they must shift to other fare.

THE MIDDLEMEN

Phytoplankton sustain possibly the most abundant kinds of animal life on Earth—10,000 or so species of copepods, mostly grazers, each individual typically no larger than a comma on this page. These crustaceans are critically important as "middlemen," translating the sun's energy as embodied in plants into their own body, swimming bits of nutritious protein, oil, and other vital substances. In the Arctic, Antarctic, and in the open sea globally, somewhat larger crustaceans, species of krill, also feed directly on phytoplankton, providing a critical link between the sun's energy and birds, fish, whales, seals, mollusks, and many other kinds of life.

In temperate waters around the world, this is where the legions of small fish come in. Most large animals in the sea are carnivores and are unable to dine directly on phytoplankton. And only a few, such as certain large sharks and whales, can support their mass by sifting zooplankton: copepods, krill, swimming snails called pteropods, and the young stages of dozens of organisms that produce floating larvae. But the small, silvery creatures collectively known as bait fish are naturally equipped to strain and convert microscopic plankton into tasty tissue.

As hatchlings, herring, like sardines, pilchards, shad, menhaden, and other clupeids, feed directly on phytoplankton, but as they grow, most consume tiny crustaceans, especially copepods, that in turn eat the phytoplankton. Other small fish have similar habits and a similar role in ocean food webs. They include the Engraulidae, with about 140 variations on the theme of anchovies and anchovetas, and the Ammodytidae, or sand lances, slim creatures that hover in huge numbers just above the seafloor but seek haven in the sand when predators approach.

Two species of menhaden, fish that consume phytoplankton even as adults, once abounded along North America's eastern seaboard and the Gulf of Mexico. They are, as a book written by Rutgers University professor H. Bruce Franklin rightly calls them, *The Most Important Fish in the Sea.* I first became aware of these small fish with a giant impact on ocean systems when I was a student at the Duke University Marine Laboratory in Beaufort, North Carolina in the 1950s. Actually, months before I met one in the ocean, I smelled tons of them being rendered into oil and meal by Piggy Potter's menhaden factory—the aroma a deep, gagging, rank, oily old socks and fertilizer essence that caused me to give wide berth to the north end of town. "Smells just like money," was the explanation my neighbors gave for why they put up with having the factory so close to where they lived.

Menhaden indeed became money for many in the 19th century, as the massive schools of fish were taken and processed in cooking vats reducing them to oil and fertilizer, an endeavor that continued far into the 20th century but not the 21st. Only one company, Omega Protein, now scours the ocean for the last remaining populations of what biologists call a keystone species, a critical link in the structure of an ecosystem. Not considered edible by most people, the little fish are gustatory magnets for bluefish, mackerel, stripers, tuna, cod, and virtually any other fish with a mouth large enough to engulf one. Those with smaller mouths, including crabs and other bottom-dwellers, dine on the crumbs that drift down like manna when big fish feast. The prodigious populations of fish reported by Captain John Smith when he arrived in Chesapeake Bay four centuries ago owed their existence largely to the even greater abundance of menhaden that in turn consumed vast amounts of phytoplankton.

Biologist Lionel Walford once tried to calculate how much ocean a herring, a close menhaden relative, must engulf to keep it in calories. He observed that North Sea fishermen in 1948 could capture 58.6 tons of herring in 100 hours, adding: "To collect plankton equal to that quantity of herring, it would be necessary to strain over 57.5 million tons of water! Indeed, the herring must do much more than that. They work very hard at it and it takes three or four years of feeding before they come to useful size."

TOP PREDATORS

When we extract millions of tons of herring, menhaden, anchovies, sand eels, or other small schooling fish from ocean food chains, we are competing with larger fish that have no choice about what to eat. Among those affected are the widely distributed, high-speed, high-energy, highly sought after animals known as tunas. Long before I met a live tuna, I was well acquainted with tuna in cans, salads, sandwiches, and casseroles, and had even been coaxed into trying thinly sliced pieces of uncooked tuna at an upscale Japanese restaurant in Los Angeles, years before sushi and sashimi became commonplace globally.

Only twice have I had face-to-face chance encounters with wild, adult tunas. The first was in 1975 while diving next to the steep wall along the outer rim of the reefs surrounding Chuuk Lagoon in the South Pacific Ocean, and years later, on the other side of the planet in the Indian Ocean, along a similar deepwater edge to Astove Island. Each time, the great fish, flashing silver-blue, were curious enough about the strange primate in their shimmering world to change course and glide by for a passing glance—but wary enough to slip away smoothly, dark shadows merging with blue infinity.

My casual interest in tunas changed to intense concern in 1990 owing to a paper placed on my desk soon after I had settled in as chief scientist of the National Oceanic and Atmospheric Administration (NOAA)—the U.S. agency that includes not only the Weather Service and U.S. environmental satellites, but also the National Marine Fisheries Service, the government agency charged with looking after and regulating use of the nation's living marine resources. In matter-of-fact terms, the paper gave stunning statistics affirming that bluefin tuna in the North Atlantic had declined 90 percent in 20 years! Until that time, I hadn't known the magnitude of the loss of the iconic "fastest fish in the sea," although I was keenly aware of the dramatic drops in numerous other ocean species worldwide.

High overall catches had been maintained in the latter part of the 20th century by the advent of industrial fishing. Fishermen were now using increasing numbers of large, fast boats, employing greatly improved methods for finding and catching various species, moving farther offshore, working in deeper water, and developing new markets for fish previously regarded ignominiously as trash fish. These included most kinds of sharks, rays, and toothy anglerfish, now a popular culinary specialty marketed as monkfish. Deep-sea fish from Antarctic waters known to scientists as Patagonian toothfish became "Chilean sea bass," slimeheads emerged as "orange roughy," beautiful, dark-eyed brotulid rat-tails were magically repackaged as "hoki," and hundred-year-old fish that can withstand freezing temperatures owing to an unusual kind of "antifreeze" in their blood came to market as Antarctic cod.

Whatever name is given to the pieces of wild fish sold as fish sticks, a fish sandwich, fish and chips, or catch of the day, what you think you are eating and what is actually being served may be like asking for chicken salad and getting chopped pelican. New methods for detecting the true nature of fish sold as red snapper, as well as other popular fish, have revealed a surprising number of different species. But who asks? And who cares?

A few weeks after I learned at NOAA about the plight of bluefin tuna, I attended a meeting where quotas for catching bluefins were being discussed and heard serious proposals to allow U.S. fishermen to continue taking tons of bluefins. Dumbfounded, I blurted out, "Are we trying to exterminate them? If so, great job! Only 10 percent left to go!"

The lingering 20th-century optimism about how much of the ocean's life could be taken without causing serious damage got a sharp jolt in 2003 with the publication of a paper in *Nature* by Canadian biologists Ransom Myers and Boris Worm, summarizing ten years of research based on half a century of accumulated data. The paper, "Rapid Worldwide Depletion of Predatory Fish Communities," brought sobering news: "Industrialized fisheries typically reduced community biomass by 80% within 15 years of exploitation. Compensatory increases in fast-growing species were observed, but often reversed within a decade. . . . large predatory fish biomass today is only about 10% of pre-industrial levels. We conclude that declines of large predators in coastal regions have extended throughout the global ocean, with potentially serious consequences for ecosystems."

Myers commented in an interview published by SeaWeb, "From giant blue marlin to mighty bluefin tuna, and from tropical groupers to Antarctic cod, industrial fishing has scoured the

Better alive: This wrasse at Cocos Island is part of the healthy reef.

global ocean. There is no blue frontier left. Since 1950, with the onset of industrialized fisheries, we have rapidly reduced the resource base to less than 10%—not just in some areas, not just for some stocks, but for entire communities of these large fish species from the tropics to the poles."

Co-author Worm continued, "Their depletion not only threatens the future of these fish and the fishers that depend on them, it could also bring about a complete re-organization of ocean ecosystems, with unknown global consequences."

The drastic declines documented by Myers and Worm should have been obvious. It should also have been a cause for grave concern that the damage to ocean ecosystems through massively destructive fishing methods was doing more than undermining future catches. It was tearing at the heart of planetary processes, vital not only for life in the sea, but for all life on Earth. Most people don't know how a computer works, but they would most likely acknowledge that every little gizmo has a function, and that pulling out apparently useless components can quickly lead to major malfunctions.

THE INCIDENTAL DESTRUCTION OF "BYCATCH"

Another jolt concerning the impact of 20th-century fishing came with the publication in 2004 of an FAO report by biologist Dayton L. Alverson and colleagues on bycatch, the taking of nontargeted creatures that get in the way of nets or are attracted to bait intended to lure other species. (The data were tallied by the fishermen themselves.) Scenes from the popular film *Forrest Gump* show how for every bushel of shrimp extracted, nets kill more than a hundred bushels of fish, sea grass, sponges, sea stars, and other sea life. Most devastating are bottom draggers or trawls that scrape the ocean floor, taking everything in their path, but according to the Alverson report, every kind of fishing gear exacts a terrible cost in terms of nontargeted victims. Data and information assembled by the World Wildlife Fund indicate that more than 300,000 marine mammals, hundreds of thousands of sea turtles and seabirds, and millions of tons of fish and invertebrates are taken as bycatch every year.

By their very nature, nets and hooks catch indiscriminately. A biblical reference in Matthew 13:47-48 points out that "a fisherman . . . casts a net into the water and gathers in fish of every kind, valuable and worthless. When the net is full, he drags it up onto the beach . . . and sorts out the edible ones into crates and throws the others away." So fish can be "worthless" and "thrown away" with impunity, but Matthew 10:29 refers to how even "sparrows falling to the ground" are noticed. Our indifference to the fate of fish in the sea relative to their winged counterparts in the sky may in part be owing to the fact that birds have always gotten better press.

Nets lost or tossed in the ocean continue fishing as "ghost nets" or traps or lines, compounding the incidental catch, sometimes for decades. In the northern Gulf of Mexico, while piloting the *Deep Worker* submersible, biologist Douglas Weaver came across an abandoned fish trap at a depth of 100 meters (330 feet). Following his directions, I found the sturdy wire container on a follow-up dive and saw several frightened fish—a blue angel, a triggerfish, a small grouper—pushing against the fine mesh, seeking a way out. A deep halo of pale bones spilling around the edges of the trap marked the fate of fish that had preceded them.

It isn't that fishermen *want* to catch turtles, seabirds, and other nontargeted species. Most regard the unwanted creatures with cool indifference; a few react with fury over the time and cost of dealing with bycatch. In her book *The Hungry Ocean,* fishing boat captain Linda Greenlaw describes how the crew of the *Hannah Boden,* discouraged by the lack of swordfish and the abundance of sharks, hung a seven-foot blue shark by a noose from the rigging, "where it thrashed wildly. Swaddled in rags saturated with lighter fluid, the shark was slashed at with knives and poked with gaffs like a giant piñata until a single match set it ablaze."

A PREDATOR BECOMES PREY

Given their intrinsic beauty and vital importance to the sea, it is puzzling why sharks in particular seem to inspire a special kind of fear, even loathing, among some who single them out for punishment. In 1975, the year that great white sharks were portrayed as evil, bloodthirsty people-eating villains in the blockbuster film

Jaws, I watched in helpless horror while more than a hundred hammerhead sharks were clubbed to death as they came on deck during a longline swordfish research survey near Cape Hatteras, North Carolina. Having spent thousands of hours underwater, I knew what all divers know—humans are not naturally on any sharks' menu! That same year, on a reef in the Coral Sea, I swam in a milling circle of more than a hundred gray reef sharks, feeling much safer than I do while driving on a freeway with cars heading in my direction at high speed, separated only by a line of yellow paint and a mutual desire to live.

Sharks can be dangerous if provoked, such as the young nurse shark, a typically calm, cool, imperturbable species, that was grabbed by the tail while resting under a ledge in Sarasota, Florida, in the 1960s. The shark put up with it for a while, but on the fourth yank whipped around and grabbed the tail of the deserving diver. The human survived the "attack," but unfortunately, the shark did not. If sharks meant us harm, there would be many more than the 50 or so bites they deliver annually to people who, after all, are recent arrivals in a realm occupied by sharks for 400 million years. Fewer than ten people a year die from shark bites while worldwide, between 20 million and 100 million sharks succumb to the "bite" of people as delivered by hooks, nets, and traps devised to catch and kill sharks for entertainment (sport), or spite (by those who believe sharks compete with them for other kinds of fish), but mostly for market. Hundreds of millions of years of survival skills have not prepared the ocean's "ultimate predators" for the planet's *ultissimo* predator.

Early in the 20th century, sharks were taken in modest numbers for their skin and oil-rich liver, but even then fishermen complained that it was possible to quickly "fish them out" in coastal areas. Later, a demand developed for "macho meat" (mako, thresher, and blue shark) in supermarkets and on upscale restaurant menus located far from any ocean, as well as for cartilage (peddled in pills as an unfounded cure for cancer). Sharks reproduce in small numbers, some yielding a pup or two every other year, others as many as a few dozen; they take years to mature and they live for decades. Despite their long and distinguished past, they are vulnerable to loss in a geological moment.

In 1980, designated as the "Year of the Ocean" in the United States, a perverse but well-intended campaign was initiated at NOAA to help fishermen by developing markets for sharks as "underutilized species" and by fostering new connections to Asian consumers. In two decades, fears about man-eating sharks shifted to *man* . . . eating sharks.

For centuries, soup made from the fins of sharks has been a traditional but rare treat in China, the primary attraction being the difficulty of obtaining the vital ingredients. Only emperors and, over time, a few elite diners had the privilege of consuming such a hard-to-come-by delicacy, one that like many other exotica developed an unearned reputation for enhancing one's virility. By the end of the 20th century, however, new wealth in Asia and new means of finding, catching, and marketing sharks made shark-fin soup much more commonplace. Although pricey, the dish began to be served at weddings, banquets, and restaurants all over the world as a gesture of respect and honor to guests. The enormous demand for shark fins—not the whole animals—initiated the barbarically efficient technique of catching sharks on longlines, hauling them on deck, slicing off fins and tail, then dumping the still living animal back to the sea.

LEARNING FROM THE ULTIMATE FISH

The year following my personal wake-up call at NOAA concerning bluefin tuna, Douglas Whynott cruised among tuna boats operating offshore near Provincetown, Massachusetts, watching humpback whales feeding on small fish. "It seemed odd that the whales were no longer the prey," he commented in his book *Giant Bluefin.* "Time, progress, had passed them by. They were no longer the source of food and light and bone. Only two decades earlier, the bluefin tuna, the 'horse mackerel,' was worth five cents a pound and served as cat food. Now this prime sushi fish was airfreighted, fresh on ice, to Japan." In the 1990s, bluefins were auctioned in the Tokyo fish market Tsukiji for more than $100,000 for a 200-pound fish. A few raw slivers served as sashimi might go for $100. High prices mean profitability even with a small catch.

While some extol the exquisite taste of bluefins, others write lovingly about their streamlined beauty and grace, notably Carl Safina, in his *Song for the Blue Ocean,* and Richard Ellis, with *Tuna, A Love Story.* Bruce Collette, scientist and tuna scholar, typically regards fish with cool, scientific detachment, but borders on the poetic when talking about bluefins, "the ultimate fish." Barbara Block, Stanford professor and tuna expert, is an unabashed admirer of the three species of bluefins and their long-distance migrations, determined by implanting acoustic tags in selected individuals. I have listened to engineers at the Massachusetts Institute of Technology (MIT) sigh with envy watching films of tunas in action, then watched them do their utmost to emulate with robotic devices what comes naturally to the tunas.

At a towing tank at MIT's Department of Ocean Engineering in 1993, for instance, I peered over David Barrett's shoulder when he was working on "RoboTuna," a sleek replica developed to figure out how the fish succeeded in capturing about 97 percent of the energy generated by the little whirlpools—vortices—formed when the tuna's tail moved back and forth. At stake are possible applications for more efficient propulsion systems for submarines.

Other treasures will forever be lost if bluefins and other tunas continue on their current slide into oblivion, not the least being answers to questions we do not yet know enough to ask. The knowledge we could gain would seem to be of greater worth than can be measured in pounds or ounces of meat. How is it possible for them to power their way across entire ocean basins to specific locations, repeatedly, without a road map? What senses enable them to locate sources of food? How do they maintain position in amazing V-shaped formations as they move near the ocean's surface? Who leads, and why? How do they rest? What incredible adventures must a hatching tuna go through during those most vulnerable first days, weeks, months, and years when thousands of hungry mouths seek to eat them, and while they must forage their way through the bewildering eat-and-be-eaten array of creatures that they meet along the way?

Of the millions of eggs released by a single adult female, a few fortunate survivors may live to reproduce themselves. Now that the number of adults has been reduced to a fraction of the population of half a century ago, will there be enough in a spawning aggregation to produce sufficient fertilized eggs to yield the necessary number of youngsters to work their way through ocean food chains to become a reproducing adult? How many eggs will go unfertilized if the number of spawners is small?

There are hundreds of recipes about how to prepare tuna raw, roasted, steamed, boiled, broiled, chopped, and otherwise offered on a plate. But what does it take to make a tuna?

As with herring, menhaden, groupers, snappers, and many other fish that release gametes into the sea, a tuna begins by being one of several million fortuitously fertilized eggs, a translucent sphere rich in oil that provides food and maintains buoyancy.

Tsukiji Market, Tokyo: way station for a million tons a year of ocean life.

The minuscule plankton that feed infant hatchlings occur near the sunlit surface; sinking eggs perish. Over the years, I have looked at a lot of plankton samples, but only once, in the Gulf of Mexico, found the distinctive great, dark eyes and spiky fins of a translucent tuna infant. Severe competition ensues as the youngsters—helpless as drifting eggs and larvae—strive to find food while avoiding becoming food for something larger. Carl Safina remarks that baby tuna live as part of the ocean where "the animals, a bizarre and minute zoo of predatory invertebrates and menacingly armed fish and crustacean larvae, spend all their time consuming . . . plants or murdering each other. A more dangerous neighborhood can scarcely be imagined."

Most of the eggs are consumed, most of those that survive to become larvae are consumed, and most of those that make it through the first year are consumed. It is a miracle that any survive all the way to six-year maturity, let alone to become thirty-year-old thousand-pound ocean veterans. The realm of plankton is a cauldron where ocean chemistry is shaped by sunlight that drives photosynthesis that in turn yields energy and oxygen and provides a major part of the underpinnings of life on Earth. When tuna eggs and young go missing from the equation, what are the consequences, not just to the future of tuna, but to the fine-tuned ocean systems that have included tunas for millions of years?

No one knows.

What we do know is that by the time a tuna weighs a pound, it has consumed many pounds of tiny carnivores that in turn have eaten other carnivores or phytoplankton. Thousands of pounds of minuscule phytoplankton are invested in every pound of young tuna, but even in the midst of plenty, baby tuna would starve without minute grazing "middlemen" to gather the green and pass it along.

Knowing these things, tuna should never be consumed casually, but always with great respect, if they are to be consumed at all. Their present precarious state makes eating bluefin tuna comparable to dining on snow leopard or panda. If present trends continue, those who really love eating tuna and most other forms of ocean wildlife will have to time-travel back to the middle of the 20th century. The results of an analysis of global fisheries data by 14 eminent scientists published in *Science* in 2006 officially confirmed the conclusions of many back-of-the-envelope calculations. Lead author Boris Worm said in an interview, "Species have been disappearing from ocean ecosystems and this trend has recently been accelerating. . . . If the long-term trend continues, all fish and seafood species are projected to collapse within my lifetime—by 2048." But, he continued, "The good news is that it is not too late to turn things around."

So, should we race to see how quickly we can consume the last tuna, swordfish, and grouper? Or race to see what can be done to protect what remains? For now, there is still a choice.

Food for thought: Octopuses are noted for their intelligence.

TAKING WILDLIFE III—THE SHELLFISH

I suppose that when the sapid and slippery morsel—which is and is gone like a flash of gustatory summer lightning—glides along the palate, few people imagine that they are swallowing a piece of machinery (and going machinery too) greatly more complicated than a watch.

—Thomas Huxley, *Oysters and the Oyster Question,* 1857

As hunter-gatherers, it is in our nature to wonder, first, if a creature we encounter for the first time might eat *us,* and second, if we might eat *it.* In my backyard explorations as a budding two-year-old scientist, I discovered how earthworms taste before I learned anything else about them, an inquisitive habit that almost killed me years later when I experimented with castor beans.

Curiosity and hunger probably inspired the first person to dare to crack open an oyster's paired shells and eat the slick gray lump within. The enormous mounds of ancient shells bordering modern and submerged coastlines globally, from China to North America, Africa to northern Europe, are tangible evidence that the first taste led to many more, ultimately leading to the enduring one-way love affair people have for *Ostrea edulis* (the

European epicurean favorite), *Crassostrea virginica* (the western Atlantic counterpart) and other variations within the family of Ostreidae mollusks that thrive in temperate and tropical coastal waters worldwide. Members of a distantly related family, Pteriidae, also called oysters, are rarely eaten but for centuries have been exploited for their lustrous shells and occasional pearls.

THE INDISPENSABLE MOLLUSK

It is a precarious life, being an oyster. Of the hundreds of millions of eggs produced in a season by a healthy young female, only some are fertilized by a male oyster's sperm, simultaneously broadcast into the open sea. Those that are fertilized are transformed within hours into translucent top-shaped creatures, trochophores, with two rapidly beating bands of cilia around their midsection. Within a day, these morph into veliger larvae, slightly larger specks of protoplasm armed with mouth, gullet, stomach, and two wispy shells. Eyespots develop, features lost as stuck-in-one-place adults but handy for youngsters surviving in a free-swimming mode.

A fraction of one percent make it through the ocean's throngs of hungry mouths to settle out as spat, or baby oysters. Once in place on a rock, piling, boat bottom, or, ideally, a handy oyster shell, growth is relatively slow—it takes three or four years to reach seven centimeters (three inches) in cool New England waters, or as little as 20 months farther south along the warmer Atlantic seaboard and into the Gulf of Mexico. Settling down where oysters already live gives a young oyster the advantage of being where others of its kind are successfully making a living, and when it comes to having offspring, if you are nestled into a thriving community of oysters, it is much more likely that your eggs or sperm will be cast into the vicinity of receptive partners.

Once anchored, oysters begin the serious business of engulfing the movable feast brought their way by tidal currents. Most live within the intertidal zone, but they can prosper in depths up to about 10 meters (33 feet), a range that encompasses sunlight-driven food production near the surface and increased numbers of predators in greater depths. Like miniature aquatic vacuum cleaners, oysters suck water in, retain most of the floating particles, and squirt out cleaner water. Every spoonful of healthy bay water they consume may contain hundreds of millions of bacteria representing more than a thousand distinctive varieties; dozens of variations on the theme of living motes, collectively known as plankton; and countless particles of organic and inorganic matter. Little green things—phytoplankton—provide the oysters' main sustenance, but other organisms are consumed as well, from non-photosynthetic bacteria to the minute larval stages of numerous kinds of marine animals. As grazers, oysters are low on the food chain, equivalent to rabbits, deer, or cows, but as consumers of the young of other animals, they jump to the next level—on a par, say, with bears.

Fortunately, *Crassostrea* can tolerate a wide range of circumstances: temperatures below freezing in winter and above 35°C (95°F) in the summer, as well as salinity ranging from an oceanic 35 parts per thousand to a dilute 5 parts per thousand at river mouths. American oysters, like their counterparts in Europe, western North America, Japan, and elsewhere, have thrived for millions of years, surviving hazards that included a 100-meter (330-foot) rise in sea level. Fossil oyster reefs lining long-submerged deepwater shorelines miles offshore from the Chesapeake Bay and from broad stretches of the Gulf of Mexico provide tangible evidence of oysters' toughness and resilience through major climatic and geographic upheavals. As times changed, they migrated upward, new colonies keeping pace with sea-level rise, while those left behind perished.

A TASTE FOR OYSTERS

Along the Hudson Valley in New York, some of my long-ago ancestors on my mother's side of the family, the Lenape Indians, consumed enough oysters starting 3,000 years ago to leave great mounds of discarded shells. When human numbers were small and oyster numbers large, bites out of the oyster's empire diminished the overall oyster population but did not eliminate enough to cause entire reefs and their associated communities of life to disappear. The shift from "sustainable" consumption of *C. virginica* to inexorable decline had its beginnings with the arrival of Europeans in North America in the 1600s, a pattern reflected elsewhere. More people = fewer oysters.

In North America, new settlers brought with them a growing number of people to feed; a taste for oysters; ingenious new strategies for extracting, preserving, and marketing them; and unprecedented ways to alter the physical and chemical nature of the coasts where oysters thrived. The biological strategy for survival that worked for *Crassostrea* over the ages began to falter.

Reports from Jamestown settlers in the 1600s and tangible evidence in coastal Indian "kitchen middens"—ancient garbage disposal sites—confirm that some lucky oysters lived for decades, occasionally attaining the size of dinner plates. Getting that old and that big means surviving not only hungry humans but also the appetites of several kinds of carnivorous snails, especially whelks and the aptly named oyster drills, as well as certain birds, notably the oystercatcher, whose bladelike beak resembles the sturdy, flat knives wielded by oyster shuckers in the Chesapeake Bay.

Sea stars, crabs, parasitic worms, shell-boring sponges, certain viruses and bacteria, storms, and climatic shifts have also tempered the oysters' natural propensity to populate the world. Enough have survived, though, to be a significant force in shaping and stabilizing coastlines from Canada to Mexico, Sydney to Singapore. Moreover, thousands of other species have come to rely on the intricate slants and curves of their hard, clustered shells for shelter.

Like coral reefs, oyster reefs provide the bricks and mortar, buildings and walls for thriving metropolises that include fish, polychaete worms, flatworms, spoonworms, nematode worms, peanut worms, anemones, sea stars, brittle stars, sponges, bryozoans, hydrozoans, amphipods, isopods, shrimp, and many kinds of crabs. Not to mention the microbes that easily outnumber in diversity and mass all the other life-forms combined. More than 15 phyla—the major genetic categories of animal life—can live within an arm's embrace on an oyster reef, about the same number of animal phyla that occupy all terrestrial environments, including lush rain forests. Take away the oysters, and a great unraveling follows.

So large and numerous were the oyster banks in the Chesapeake Bay when Captain John Smith arrived in 1607 that they were described as hazards to navigation. Tasty, too, it appears. Although the population of native people and new arrivals numbered fewer than 30,000 by 1650, within a hundred years, more than 300,000 lived near the bay, many deriving their primary nutrition from shellfish and other local wildlife. Clearing the land of forests for agriculture provided important new sources of farmed food but resulted in the loss of so much soil into the bay that navigation was inhibited and thousands of acres of coastal wetlands and prime oyster habitat were destroyed. By the late 1600s, relying on wild birds, deer, squirrels, and other terrestrial wildlife for food and trade in feathers, hides, and furs had largely given way to an economy based on agriculture, but the serious taking of aquatic wildlife for commerce was just beginning.

In 1700, New York City's population numbered only about 5,000 and oysters numbered in the millions, but a mighty shift was about to take place. By 1800, there were more than 60,000 residents of New York. Hundreds of thousands of oysters were being consumed locally and hundreds of thousands more exported elsewhere in the U.S. and the Caribbean, pickled. Half a century later, the human population had risen to more than half a million in the city, and new European markets were being opened for pickled American oysters. Mark Kurlansky observes in *The Big Oyster* that by the beginning of the 20th century, three million New Yorkers were consuming on the order of a million oysters *a day,* and more were taken for distant markets.

Nothing in the long history of *Crassostrea* had prepared them to cope with such colossal consumptive capacity. But primate predation was just part of the problem for oysters and the millions of creatures that depended on them for food or shelter. Sewage flowing freely from the city and local farms increasingly contaminated coastal waters, along with tons of silt and debris from land shorn of its native vegetation. Even so, it is estimated that every few days, the entire contents of New York Harbor were filtered through the prodigious number of oysters living there at the time, each small outgoing gush purer than those incoming. Farther south, all of the waters of the Chesapeake Bay may have passed through the bay's oysters and clams each 24 hours, an estimate based on the numbers of mollusks thought to populate the area at the start of the 20th century and their known capacity to pump water. Fewer than 2 percent of the oysters, clams, crabs, and sponges—and menhaden—that once prospered in the bay now remain to cope with present loads of silt, sewage, and algae.

By the time I came along as a child in the 1940s, great dents had already been made in North America's ocean wildlife. Nonetheless, hunting and gathering the wild creatures that lived in the marshes and inlets bordering the western edge of the Atlantic Ocean was still a way of life for many. Watermen—the aquatic version of terrestrial hunters who seek, capture, and market wild creatures—did not have to travel far to find and gather ducks, geese, blue crabs, terrapins, striped bass, bluefish, flounder, clams, and, of course, oysters. I know what New Jersey skies look like darkened with wave after wave of migrating songbirds, know the look and sound of tens of thousands of wild ducks and geese descending with chaotic grace into wild marshes where one of my uncles made a living as a market hunter. I know the thrill of being a hunter myself, nine years old, lowering a scrap of meat on a weighted line into the clear, dark water in a marshy back channel where the pale claws of blue crabs magically appeared to clutch the bait.

Every Friday, a truck with large, spoked wheels rattled down the lane leading to my family's New Jersey farmhouse, ringing a bell that added a few high notes to the background clatter of the fish man's arrival. White-fleshed flounder, deep orange shad roe, blue crabs glowering and snapping in glistening piles, flinty-shelled oysters, and sometimes, my Dad's favorite, cherrystone clams, buried in heaps of glittering ice—all were caught by the fish man's suppliers, men who worked along the eastern Jersey shore or somewhere in the shallows of the Delaware Bay.

Oyster stew meant a meal of salty gray mounds floating in steaming bowls of hot milk, a swirl of butter, a dash of pepper, a handful of hard, round oyster crackers, and a search to see who could find the small pink pinnotherid crabs, cooked with the oysters and treasured as dividends by my brothers and me. Not parasites, just tiny lodgers known as pea crabs that as planktonic juveniles are swept as minuscule flecks into the space between the oyster's shell and its soft tissue, there growing to fingernail size, sharing space with their molluscan host. The crabs settle down for life within the oyster's silken folds, the males eventually leaving on a one-way journey to find and fertilize the eggs of females living in nearby shells. Hatched eggs become exquisitely small shrimplike beings, most contributing their minute morsel to the great ocean food web. A lucky few find oysters of their own and continue the cycle, specks among the myriad small, medium, and very large creatures that individually and together harmonize as an ecosystem that until the most recent fractional blink of time, geologically or biologically or astronomically or even *gastro*nomically speaking, has not included humankind.

SINGLE-SHELLED BEAUTIES

Of the 100,000 or so kinds of mollusks believed to exist currently, a few dozen are in the family that includes edible oysters. Other bivalves that people eat range from the giant "clam" famous in the South Pacific to the hard-shell clams in clam chowder, the razor clams typically sold fried, as well as various kinds of scallops, mussels, cockles, and even the tiny coquinas that inhabit certain tropical sandy beaches. A number of mollusks with a single shell, the gastropods, also have become human fare, notably various species of abalone. People and abalones seemed to get along pretty well for thousands of years in coastal areas of western North America, as well as in Australia, New Zealand, Japan, and South Africa, where large species of abalones naturally occur. But as in other scenarios, when the number of wild animals taken exceeded the ability of the creatures to replace themselves, trouble followed. In California, starting in the 20th century, enormous piles of abalone shells marked commercial use of abalones on an unprecedented scale, mostly for food, but also supporting a secondary industry—buttons crafted from the shells' shining, iridescent lining.

By taking mainly the big, adult abalones, people extract the fortunate survivors of a rigorous selection process that begins

Clams such as these are part of the ocean's natural filtration system.

when hatching abalones join a perilous planktonic free-for-all where they are hunted by hundreds of other species. Once settled down as bottom dwellers, the dome-shaped creatures plow along like small tanks, grazing on seaweed, tender sprouts of red algae, and young blades of kelp. It takes about eight years for them to mature, and lucky ones may live for decades.

Curiously, as abalones began steadily, then sharply, declining in California waters in the 20th century, a modest recovery began for sea otters. Those sleek, furry mammals lived compatibly in kelp forests with abalones for about five million years, including the recent 10,000 or so when people first began populating coastal areas in North America. Originally numbering about

300,000, most of the sea otters were killed by hunters for the fur trade between the mid-1600s and 1911, when they were given full protection in U.S. waters. Abalones were relatively plentiful when a relic population of about 100 otters was discovered off the California coast in 1938. Now abalones are scarce, while nearly 3,000 otters currently occur along the central California coast. Some see a cause-and-effect correlation—more otters, fewer abalones, discounting the impact of commercial abalone hunting. No doubt about it, otters eat abalones, as well as crabs, clams, snails, slugs, urchins, sea stars, worms—and numerous other creatures that stoke their hot-blooded metabolism.

But ecologists suggest that otters are, in fact, an abalone's best friend: Otters help to keep down the population of sea urchins, which compete with abalones for food. Sea urchins, like abalones, graze on seaweed, but the urchins can reproduce more quickly, and not only consume young kelp, but chew through large, grown-up holdfasts and tough stalks, essentially mowing down the forest. In an otter-free ocean, urchins get a lift with the loss of one of their major predators, and abalones take a hit with the loss of kelp. In a healthy ecosystem, there is a place for everybody—kelp, otters, abalone, urchins, even a few people along with thousands of other dynamic, living pieces. The good news is that with better understanding of the system, coupled with care, recovery can follow. The proportion of ingredients is somewhat out of whack, but the elements are still there. For now.

The same is true in the realm of another kind of big, tasty snail, the pink queen conch *Strombus gigas,* served as food, carved into ornaments, and so important to the economy in the Bahamas and Florida Keys in the 1800s that longtime residents there became known as conchs—an epithet borne with pride. Change for the queen conchs came with unsustainable taking, evident by 1900 and continuing to the present time.

As with oysters, abalones, and most other marine mollusks, conchs spend the first part of their life playing eat-and-be-eaten plankton roulette. Those that survive to begin life as minuscule-shelled bottom crawlers face another round of predatory obstacles: crabs, fish, and lobsters. The few conchs that survive the first two or three years may keep going for two or three decades. It takes at least five years for them to begin to develop the characteristic flared lip valued by those who treasure them as mantelpiece ornaments, the base of a lamp, or the setting for a clock.

The walls, driveways, and heaps of empty *Strombus* shells in parts of the Caribbean are stark reminders of the millions of plates and bowls of conch fritters, conch ceviche, cracked conch, and conch chowder that not only provided calories to people who lived nearby but also yielded income when sold commercially to growing crowds of visitors as well as to distant markets. They are now protected in the Florida Keys and parts of the Caribbean, and I was impressed recently with how concerned officials in Belize are about restoring the big pink snails to their waters. Inspired by Birgit Winning, founder of the Oceanic Society's field station on Blackbird Caye, Belize, some of us on a dive trip decided to help the local economy and the future of pink conchs by initiating a "buy and release" effort. Pooling resources, we bought dozens from the still living conch catch of local fishermen and transported them to sea grass meadows in a nearby "no-take" reserve. Rather than just pitching them over the side, each conch was hand-delivered to a safe haven below, a procedure that looked so fishy that patrolling park officials came racing over. Only Birgit's credibility and diplomatic skills—*"Really, officer, we were putting them IN the reserve!"*—saved us. The real trick, though, will be saving the conchs for good. Their continued appearance on menus on cruise ships, in restaurants in Florida and throughout the Caribbean, parallels their disappearance throughout most of their natural range.

THE CLEVER CEPHALOPODS

Within the molluscan tribe are some so endowed with superior craftiness, curiosity, and athletic skills that scientist-poet Loren Eiseley once remarked that it is "just as well for [humankind] that they never came ashore." He was referring to cephalopods: octopuses, squids, cuttlefish, and nautiloids, all creatures with exceptional eyes, wits, and an appetite for meat—there are no plant-eating cephalopods. The importance of their role in ocean ecosystems is incalculable. Squids, for instance, occur in enormous schools, dining on smaller prey and becoming prey in turn for creatures that are adapted primarily to consume them.

The celebrated squid of all squids, *Architeuthis dux,* is believed to be the largest, heftiest invertebrate on Earth. With a body the size of a bus, eyes bigger than soccer balls, and arms that may extend for more than the length of its body, *Architeuthis* is a magnificent animal, no matter what your perspective. "Calamari for a month," said a visitor gazing at the 8-meter (27-foot) specimen on display at the Mote Marine Laboratory in Sarasota, Florida. "A worthy meal for deep-diving sperm whales," commented Greg Marshall, who has suction-cupped cameras to the backs of whales, hoping to film the legendary mollusk-mammal encounter. "Bliss," according to the Smithsonian Institution's cephalopod specialist, Clyde Roper. "The ultissimo squid!" said Mike deGruy, who has observed and filmed more cephalopods in the wild and in laboratories doing more outrageous things to more mazes and puzzles than any other human.

I have been the object of intense curiosity expressed by a lustrous, silvery red octopus while sitting otherwise alone in a little submarine, 396 meters (1,300 feet) under the surface of a stormy sea off the coast of Lanai, Hawaii. Mistaking her at first for a big piece of floating trash, I started to move away, then saw the eyes. When I turned the sub in her direction, she backed off a little, but I waited, and she came closer. I moved the sub to get another angle, and she moved with me. For an hour we danced upward more than 305 meters (1,000 feet) in the water column, each eyeing the other, each able to pull away but neither exercising the option, until finally, reluctantly, I had to return to the surface.

Fifteen meters (50 feet) underwater in a cove near Sydney Harbor, a lacy, gold-speckled cuttlefish the size of a serving platter went out of her way to come to where I had stopped to inspect a leafy clump of kelp. I was already lying flat on the bottom but shifted to face the ten probing arms reaching in my direction. In return, I extended my ten fingers, and for some minutes we simply exchanged touches and close scrutiny. Eventually I had to leave owing to a diminishing air supply, and as I lifted off, she glided into the sheltering kelp, leaving me wondering whether or not she might also have the capacity for wonder. An encounter with an alien from another planet as inquisitive and intrinsically beautiful as that cuttlefish would have made headlines, something I think about when I see stacks of her cousins, fresh, frozen, and sometimes dried and nicely seasoned, in Asian markets.

To albatross, winged giants that can stay aloft above the open sea for a year, small, oceanic squid provide vital sustenance; this is true, too, of legions of other seabirds, fish, and mammals that obtain their share of the sunlight's energy transformed through food webs via squid. Some squid have become moneymaking tourist attractions, notably the thumb-size "firefly squid," creatures studded with light-emitting photophores. When trapped in a net, the entire body of each little squid glows a fiery indigo, and luminous ink lights the surrounding sea with clouds of blue fire. Anyone seeing only their stiff gray bodies as they appear in the market, neatly arranged like cookies in a tin, could never reconcile that image with the vision of them alive, arrows of blue lightning in a dark sea.

Nearly three million tons of squid are taken annually, usually by attracting them with powerful lights, then using nets or mechanized reels of hooks that snag the squid and lift them to bins on the decks of surface ships. The rich, highly productive waters at the edge of Argentina's Patagonian shelf attract so many squid boats that illumination from their lights rivals major South American cities when viewed from space. In Monterey, California, lights from the squid boats serve as a beacon for miles around, disrupting night rhythms under the sea and in the skies above, while luring millions of creatures, gathered together to spawn, to a destiny on a plate—or as bait for pricier wild creatures.

Three hundred or so different kinds of octopuses and squids grace the ocean, their ancestry apparent in the fossil record extending back more than 500 million years. Dinosaurs were 300 million years in the future. The DNA that makes us special wouldn't come along for nearly half a billion years after cephalopods! Looking into the eyes of an octopus or squid or their frilly-skirted cousins, the cuttlefish, I get the uncanny feeling that somebody is home there, somebody who is regarding me with more than casual interest. Their value as ambassadors from the past, loaded with valuable insight concerning the nature of life on Earth, would seem to far outweigh their worth as bait, food for livestock, or as an occasional meal. Yet, throughout the 20th century, few have expressed concern about the consequences of large-scale taking of these ancient creatures, and there is not a large "save the cephalopod" constituency entering the 21st.

CRUSTACEANS BIG AND SMALL

Similarly, extraction of enormous numbers of crustaceans from the sea—crabs, lobsters, shrimp, and krill, with a deep ancestry similar to mollusks'—arouses little concern except from those who fret that the cost of their favorite seafood might rise or availability decline. Sixty thousand or so kinds of crustaceans are included within the most diverse group of animals on Earth, the Arthropoda. Terrestrial counterparts are insects, centipedes, spiders, and other "joint-legged" animals, and no one knows how many species of them exist, from low estimates of a million to more than 30 million. Most of them go about the business of keeping the planet functioning, one beetle, bee, ant, amphipod, isopod, krill, and crab at a time, mostly unnoticed, and largely unloved by humans, who are the unwitting beneficiaries of their existence.

Many crustaceans, like squid, are endowed with the ability to emit intense, blue-green bursts of light, useful in the deep sea, where it is dark below 305 meters (1,000 feet) or so, and in Antarctica, where sunlight is limited to half the year. Technically, no one owns the Antarctic continent, but by international treaty the land and surrounding ice shelves have been protected since 1961, forestalling commercial exploitation. What nations do or don't do in Antarctic waters is governed by a subset of the Antarctic Treaty, the Convention on the Conservation of Antarctic Marine Living Resources (CCAMLR). The stated objective of CCAMLR is "to safeguard the environment and protect the integrity of the ecosystem of the seas surrounding Antarctica, and to conserve Antarctic marine living resources." That has been interpreted to mean that it is all right to take millions of tons of krill, with consequences that are yet to be determined. Of all the creatures in Antarctic waters, the reddish, shrimplike *Euphausia superba* serve as a critical linchpin; tug on them, and the entire system moves.

Justification for taking krill from Antarctica was explained in the 1980s as a logical effort to correct an "overabundance" brought on by the lack of krill-eating whales, depleted by earlier overexploitation. Dozens of fish species, penguins, terns, squids, and the remaining seals, sea lions, and whales may have had a boost in the number of krill available to them with fewer whales, and it isn't likely that any of them would have regarded this as a problem, let alone one that required a fix from factory trawlers from the U.S.S.R., Japan, Chile, Poland, South Korea, and more recently, Norway.

Among crustaceans, only a few in the category characterized by having ten legs, the decapods—lobsters, crabs, and shrimp—are coveted by humans as food, but they and the thousands of other species in the group are critically important elements in the vast living machinery that makes the ocean function. Blue crabs in the Chesapeake Bay have a place that transcends crab cakes and she-crab soup, and their swift 20th-century decline—along with the depletion of oysters, clams, sponges, and menhaden—is correlated with the decay of the bay and other coastal ecosystems throughout their range. The extraction of millions of tons of the large, spiny king crabs from Alaskan waters in the latter part of the 20th century has surely left a gap in the systems from which they were wrenched. The catch dropped 60-fold between 1980 and 1983, a consequence, some say, of a warm-water trend or "excessive predation by fish on the young crabs." But in their long history, until the 20th century, the list of things for a king crab to fear did not include primates in yellow slickers.

Decapods include a wondrous suite of tiny shrimp that have developed critically important relationships with vastly different phyla of animals. Some set up housekeeping within the silky embrace of an anemone's tentacles, immune to the stinging cells that stun or kill most small creatures. Others hang out with moray eels, surrounding their toothy hosts like crimson ornaments, dining on scraps and fish parasites. On a healthy coral reef, shrimp-cleaning stations are well known to resident fish, often resulting in traffic jams as groupers, jacks, parrot fish, and others mill about, waiting their turn to be groomed. Wounds have bits of flesh trimmed, parasites are clipped away. I am always somehow pleased but disconcerted when cleaners hop on my arm and use their tiny forcep-like claws to preen individual hairs that I thought were already pretty clean. Once, I watched a diver remove his mouthpiece to allow several eager shrimp to hop onto his lip and work over the spaces between his teeth.

There is an old Chinese saying that "big fish eat little fish, little fish eat shrimp, and shrimp eat mud." Shrimp don't actually dine on mud, but many species, including the several dozen kinds of shrimp and prawns that are taken commercially for human consumption, do feast on detritus—organic material that accumulates on the seafloor, where the crustacean cleanup crews make a good living as scavengers. Removing these important creatures from the sea for an endless flow of prawns boiled, fried, steamed, grilled on the barbie, in a cocktail, soup, or u-peel-'em pile seems a poor trade-off, especially when the trawls used to bring them from the sea to the surface cause permanent mayhem on the ocean floor.

Shrimp have about 30 large-clawed cousins, the lobsters, that have inspired at least as many recipes and created comparable controversies over their extraction from ocean ecosystems. One of them, the New England icon, *Homarus americanus,* the American lobster, was closely studied by biologist Francis Hobart Herrick, who observed in 1895: "We have witnessed in the lobster fishery for many years past the anomaly of a declining industry with a yearly increasing yield, but with the gradual diminution in the size of the lobsters caught and an undue increase in the number of traps and fishermen."

Herrick noted that in 1886, 90 million lobsters were taken in Canadian waters alone, principally from Nova Scotia. Since it takes newly hatched larvae a perilous five to six years to wend their way through astonishingly complex changes in shape, food habits, and habitat to reach "legal size"—about a pound—it is incredible that any, let alone so many, reach maturity. The adventures of decades-old 30- or even 40-pound giants are scarcely imaginable.

Even more mind-stretching is figuring out how ocean ecosystems functioned before the advent of humans as serious predators, and how they manage to cope now with us as a sudden major factor in their strategies for survival. There is no surplus in natural, healthy systems, just waiting for people to exploit. Whatever we remove is a loss for those who developed a taste for lobsters long before we came on the scene, and a distortion of the energy flow within the ocean overall. The ocean's investment in making 90 million lobsters in one area may seem small in the overall scheme of things, but the extraction of that number, year after year, magnified by taking comparable quantities of numerous other species over increasingly large areas of the sea will eventually, inevitably, inexorably rock the system.

Herrick fumed about this more than a century ago in his hefty tome on the American lobster:

> Civilized man is sweeping off the face of the earth one after another some of its most interesting and valuable animals, by a lack of foresight and selfish zeal. . . . If man had as ready access to the submarine fields as to the forests and plains, it is easy to imagine how much havoc he would spread. The ocean indeed seems to be as inexhaustible in its animal life as it is apparently limitless in extent and fathomless in depth, but we are apt to forget that marine animals may be as restricted in their distribution as terrestrial forms, and as nicely adjusted to their environment.

Access to the "submarine fields" grew swiftly in the 20th century, coupled with increasing numbers of people with an appetite

Missing in action: octopuses are critical components of a healthy ocean.

for lobsters and other ocean wildlife. As predicted, species all over the world that had been abundant in Herrick's time began to collapse. However, despite some major ups and downs for lobsters and lobstermen in North America, and despite greatly altered coastal ecosystems, the glossy greenish brown clawed creatures that turn shockingly orange-red when boiled remain a major part of the total wild catch of marine life in New England. The reasons for their continued prosperity relate to various conservation policies coupled with the drastic reduction in cod, halibut, seals, sea lions, sharks, and other natural predators. In effect, we have eaten the lobster-eaters, while restricting the numbers of lobsters we take.

Related, but lacking large claws, are about 40 different kinds of spiny lobsters, thorny-backed animals that live in tropical and temperate seas globally. Popularly referred to as bugs, owing to their resemblance to distant insect kin, spiny lobsters probably have been eaten by people since soon after the first encounter between human and crustacean, thousands of years ago. The number of spiny lobsters that existed in temperate and tropical waters before 1900 worldwide would astonish modern "bug-hunters." Even in the 1950s, during my first underwater explorations of the Florida Keys, it seemed that spiny lobsters were everywhere, their long antennae seeming to sprout like whiskers from ledges and crevices.

As a biology student, I learned something about the strange, translucent larvae of spiny lobsters, with their long, stalked eyes and crystalline claws, creatures that would look right at home seated at the bar among *Star Wars* aliens. After dodging death by a thousand bites for almost a year, a larval spiny lobster must next survive another two or three years in sea grass meadows and coral reefs, where it is sometimes difficult to tell what is edible from what might be eyeing you for lunch. Clams, worms, carcasses of all sorts are standard fare for growing as well as grown-up lobsters. Snails that chew on live coral are in turn chewed on by lobsters, a system that benefits all players, including the snails. Too many snails mean too little coral, or else too many corals afflicted with diseases that invade the newly gouged snail trails. It takes a while, a few thousand millennia or so, but in time, things get sorted out so there is a place in coral reef ecosystems for everyone.

Taking lobsters commercially in Florida did not seriously get under way until early in the 20th century, and it continues, though much reduced, into the 21st. California's big, robust spiny lobster counterpart once supported a major industry, but lobster scientist Herrick observed in 1911, "The California langouste . . . is often trapped in great numbers, but even twenty years ago . . . the species was in danger of extermination from overfishing." It is the same story from the Mediterranean Sea to South Africa, Australia to the remote Galápagos Islands: In a few decades, we gorged ourselves on the living wealth of the sea generated during thousands of years.

Over the years, as a member of a seafood-loving family, I learned how to wield an arsenal of picks, hammers, forks, and shell crackers to extract and savor every last salty, succulent morsel from the intricate system of pipes and wiring that holds lobsters together. But now that I know how precarious the future is for these ancient creatures, and how important they are to the health of coral reefs, I have decided to cease and desist, hoping that every lobster I do not eat will increase the chances that somewhere, a lobster might live and do what lobsters do as critical players in a healthy ocean.

Our near and distant predecessors might be forgiven for exterminating the last woolly mammoth, the ultimate dodo, the final sea cow, and the last living monk seal for lack of understanding the consequences of their actions. But who will forgive us if we fail to learn from past and present experiences, to forge new values, new relationships, a new level of respect for the natural systems that keep us alive?

A collage of modern flotsam and jetsam on a Belize beach.

THE ULTIMATE GARBAGE DISPOSAL

Let's talk trash. . . . Only we humans make waste that nature can't digest.

—Charles Moore, from a speech at the TED Conference, 2009

It's as big as Texas . . ."

"Twice as big as Texas . . ."

"Twice as big as France . . ."

"As big as the continental United States!"

Actually, no term is quite adequate to describe the dimensions of the great mass of human-made debris, the so-called Great Pacific Garbage Patch, which swirls in the currents of the North Pacific Ocean. It is connected to the entire ocean, the planet's circulatory system, and all of that ocean, Pole to Pole, is clogged with large, medium, and especially, small pieces of plastic and other goods discarded by people worldwide. Like modern archaeological confetti, tiny bits of colorful plastic get caught up with bottles, shoes, plates, buckets, forks, spoons, cups, straws, toys, toothbrushes, packing straps, shavers, lawn chairs, bottle caps, coolers, crates, bags, and much more, carried by ocean currents across the face of the planet but collected in especially high concentrations in a few favored places. The northeastern portion of the Pacific gyre is one of them, with a counterpart in the western portion. A large mass has been sighted off the coast of Chile, another along the western coast of Antarctica. The northwestern Gulf of Mexico gathers in and holds more than most, and so does the Sargasso Sea, about 7.7 million square kilometers (3 million square miles) of quiet water in the central Atlantic, famous for its floating forest of golden sargassum weed—and its strange collection of human-made flotsam and jetsam.

So how big is the garbage patch? It is as big, as wide, as deep as the ocean itself. On every dive I have made in the past 30 years, whether snorkeling or in deep-diving submarines, trash of some sort, and sometimes of many sorts, is visibly present. More than 305 meters (1,000 feet) underwater in the *Deep Rover* sub, I once spent nearly an hour getting close to what I thought was a strangely sparkling deep-sea creature, only to discover that with my cautious approach I had succeeded in not frightening away a half-buried soda can. In the deep sea and on beaches from the high Arctic shores of Norway's Svalbard Islands well north of the Arctic Circle to beaches bordering Antarctic islands far to the south, trash glitters, shines, and accumulates.

Of course, people have been throwing things into the ocean—and in heaps and piles on the land, too—for as long as there have been people. But now there are seven billion of us generating the kinds of things that do not easily melt into the landscape—or seascape. Moreover, the 20th-century idea that there is a limitless supply of oil led to the belief that plastics derived from oil are also, essentially, limitless.

ETERNAL PLASTICS

Far and away the most abundant, troublesome, persistent, deadly debris in the sea is composed of plastic. A U.S. Ocean Studies

Board (OSB) report in 2008 defines "plastics" as a term that encompasses "the wide range of synthetic polymeric materials that are characterized by their deformability and can thus be molded into a variety of three-dimensional shapes, including a variety of common materials such as polypropylene, polyethylene, polyvinyl chloride, polystyrene, nylon, and polycarbonate."

The same report defines marine debris as "any persistent, manufactured, or processed solid material that is directly or indirectly, intentionally or unintentionally, disposed of or abandoned into the marine environment." That means everything from Russian nuclear submarines abandoned in the North Sea to a plastic-foam cup tossed overboard on a weekend outing.

I know it is possible to survive without plastics, although people born since the 1970s might not think so. Plastics have slithered quietly into every aspect of human society, seducing us with their multiplicity of uses, their durability, and their transmutability, easily assuming the most mundane and the most exotic forms. I brush my teeth with a plastic toothbrush, comb my hair with a plastic comb, step into shoes with plastic soles, slip on a fleece made of *plastic*. Yet I remember a time when diapers were made only of cloth, when cups were metal or ceramic, and there was no such thing as a plastic bag. All of human civilization got along perfectly well without plastics for as long as there have been people—until the past few decades. We could do so again, but whether we possess plastic or not is not the problem. The problem is the magnitude of synthetic materials that are used briefly, then thrown away for eternity, thereby permanently changing the nature of the world.

NATURAL RECYCLING AND HUMAN CLEANUP

The bright side of all this is that healthy natural systems can, in time, soften even heavy-handed human impacts. I once found

Medical waste degrades the ocean's health.

an enterprising hermit crab with its vulnerable posterior neatly tucked into a discarded Bayer aspirin bottle, a modern, lightweight, durable substitute for traditional snail shells. A decorator crab on a nearby reef had artfully placed a disposable fast-food ketchup envelope on its back along with bits of algae, hydroids and normal camouflaging elements. The ketchup container actually helped the crab blend in with other trash.

On assignment with the National Geographic Society in 1975, I spent many weeks exploring the World War II shipwrecks of Chuuk Lagoon, documenting the transformation of tanks, planes, trucks, guns, large globe-shaped mines, cases of ammunition, and the ships themselves into hauntingly beautiful homes for parrot fish, damselfish, lionfish, angelfish, tiny gobies and large jacks—a full suite of coral reef organisms, including the corals themselves.

Knowing the precise minute when the ships settled on the seafloor, I could determine minimum growth rates for some species, as well as trace development of entire communities. The plants and animals on the ships did not just magically appear; they moved in from nearby waters that had not been blasted with bombs and saturated with diesel fuel. Kimio Aisek, dive master at the Blue Lagoon Dive Shop, who observed the action as a boy, told me, "The sky and ocean thundered with exploding bombs and shells. There were dead fish everywhere and beaches were black with oil for months." But he added, looking out over the now peaceful reefs and ocean beyond, "It has healed, now."

Healed, but scars remain there and everywhere we have waged war, lost ships, dredged channels, and dumped our trash. Glass bottles and metal cans are unsightly when tossed into the sea, but are basically inert. Cans corrode over time, and glass, while durable, appears to do no harm. Octopuses, gobies, and even young groupers adopt jars and cans as safe havens, and old bottles, encrusted with coral and algae, are sometimes retrieved centuries later as treasures. Sea glass, broken pieces with edges softened and surfaces sanded by years of tumbling on a sandy shore, is prized by beachcombers. Newly broken glass is not, however, as many barefoot swimmers have discovered, painfully.

Large pieces of trash are eyesores, and some kinds, especially plastic bags, are lethal to sea turtles, whales, and whale sharks when the indigestible material is engulfed and jams their digestive system. A whale, washed ashore in California in 2007, died of "unknown causes" but had 181 kilograms (400 pounds) of plastic in its stomach. Lost and discarded fishing gear causes major problems by entangling and killing marine mammals, birds, fish, and other marine life, as well as by endangering boat propellers and submarines. Even submersibles can be brought to a halt when snagged with masses of old fishing nets, such as the Russian sub trapped in the Bering Sea in August 2005. The seven sailors aboard were rescued after their encounter with the derelict gear; most marine mammals, turtles, birds, and fish are not.

A few entangled seals, sea lions, otters, and whales are fortunate enough to be released from the painful embrace of monofilament necklaces and nets by teams of volunteers—people who have developed the special skills necessary for dealing with big, lively animals that don't know the rescuers' intentions are benign. A special group of volunteers in Hawaii have become experts in freeing humpback whales festooned with crab traps dragged all the way across the ocean from Alaskan waters, still bearing marks identifying their place of origin.

Beach cleanup efforts now organized worldwide are helping to pick up the deadly avalanche of junk once it washes ashore, and in some coastal areas, scuba divers are being mobilized to retrieve derelict fishing gear, waterlogged plastic, and a full cross-section of everyday throwaway goods. Anything and everything, including kitchen sinks, shows up. Ocean Conservancy, a Washington-based nonprofit organization, has organized and documented annual cleanups in the United States since 1986 and internationally since 1989. On a September cleanup day in 2008, nearly 400,000 volunteers in 104 countries gathered 3 million kilograms (6.8 million pounds) of trash, mostly from the edge of the ocean, but along some inland waters as well. After all, all rivers lead to the sea, and with the rivers go the toxic loads that eventually flow into the ocean. Among the volunteers were 10,600 divers. Most carried with them cutters to snip through the miles of monofilament line and nets that smother popular fishing areas.

Forty-three categories of debris were tracked and documented, revealing something about the nature of human nature. The top ten, accounting for 83 percent of the total, were, in order of abundance:

1. Cigarette butts
2. Plastic bags
3. Food containers
4. Caps and lids
5. Plastic bottles
6. Paper bags
7. Straws and stirrers
8. Cups, plates, eating utensils
9. Glass bottles
10. Beverage cans

NURDLES, PELLETS, AND DYES

Less obvious, much harder to retrieve, and more insidious are small bits of plastic, fragments of larger objects, as well as countless tiny pale pebbles dubbed nurdles, the pre-production plastic spheres that are later melted and molded to produce thousands of products, from juice jugs and action figures to lamp shades and chairs. Over 113 billion kilograms (250 billion pounds) of nurdles are created from petrochemicals every year and transported by trucks and tankers, loaded onto containerships, and carried to global destinations. The pellets are light, bouncy, and hard to restrain and readily wash down drains, blow into rivers, or spill directly into the sea. Some are deliberately incorporated into cosmetic "scrubs" and when washed off, slide down pipes and eventually flow to the sea.

Crossing the Bay Bridge from San Francisco to Oakland on a breezy day, I once witnessed thousands of escaped plastic-foam packing pellets scurrying along with traffic, some smacking into windshields, others rolling along, keeping pace with the cars and trucks, all eventually destined for San Francisco Bay and the open sea beyond. Some may be basking now on a beach in Mexico or Fiji or Japan. Similarly, I imagine the mass migration of the miniature plastic pellets, leaking from this place or that, riding winds and waters, skipping along highways toward some part of the ultimate waste disposal site, the ocean.

One way or another, it has taken only about 40 years for nurdles and other bits of plastic to rival the grains of sand on beaches around the world. Beachcombing children call the pearly plastic spheres mermaid tears.

Some sand comes from bits of broken shell or fragments of coral and coralline algae that may accumulate fairly rapidly, in geological terms. Other sand starts as mountains that get ground down to boulders that break into rocks that become pebbles that eventually become tiny bits of basalt, granite, quartz, and other minerals, a process that may take a hundred million years or so. It is a staggering thought. Future geologists will be able to precisely mark our era as the Plasticozoic, the place in the sands of time when bits of plastic first appeared.

Thor Heyerdahl, anthropologist and ocean explorer, told me that during his famous *Kon-Tiki* expedition in 1947, he and his crew sailed over pristine South Pacific seas with no evidence for weeks that other human beings existed. However, during the 1969–1970 *Ra* travels across the Atlantic, he said, "We were never out of sight of trash drifting on the surface of the sea." Alarmed at the blobs of oily waste and debris, including some of the first evidence of drifting plastic, he alerted the United Nations and the public at large, but many were skeptical, thinking he had exaggerated what was there. Other reports followed, though, and soon it was obvious that the ocean everywhere literally was being trashed.

A report from the National Academy of Sciences published five years later reported that over 6 billion kilograms (14 billion pounds) of garbage were deliberately dumped into the sea every

A moray eel makes good use of a trashed tire in the Galápagos Islands.

year. Most came from merchant ships, but about 450 million kilograms (a billion pounds) of it was discarded from fishing vessels, with passenger vessels, recreational boats, oil drilling platforms, and other sources adding to the mix. Trash from military vessels was not included, but until the 1990s, all wastes from U.S. Navy ships were discarded into the sea—more than 450 kilograms (1,000 pounds) a day for a single large vessel.

In 1988, an international convention, MARPOL Annex V, came into force that began regulating marine debris. It prohibited disposal of plastics but allowed other garbage to be put into the sea. In the United States, actions are implemented through the Act to Prevent Pollution From Ships, a law that makes it illegal to dispose of plastics at sea and includes some restrictions on the disposal of other garbage. Making the law was one thing; enforcing it is another. At sea aboard a passenger vessel in the Indian Ocean during the 1990s, I watched crew members tossing large plastic bags filled with trash over the stern late at night. They bobbed away in the moonlight, disappearing over the horizon, but were certain to reappear on a distant beach or somewhere on the ocean floor.

Inspired by Heyerdahl and deeply concerned about ocean debris, explorer and conservationist David de Rothschild dreamed up a way to create global awareness and encourage industry to focus on win-win solutions to plastic pollution. Of the 39 billion plastic bottles that are used in the United States alone every year (about 2 million every five seconds), only about 20 percent get reused. Rather than sending a message in a bottle, de Rothschild decided to make a message *of* the bottle—thousands of them.

In 2009, he is launching *Plastiki,* an 18-meter (60-foot) catamaran built entirely of recycled plastic bottles, for an expedition from San Francisco, across the Great Pacific Garbage Patch, through islands of the South Pacific, to Sydney. A small crew, including Josian Heyerdahl, Thor's granddaughter, will document the voyage, sending a stream of live state-of-the-ocean reports to watchers worldwide. Rather than wind up in a museum or landfill, *Plastiki* is destined for recycling, perhaps as fleeces that eventually can be turned into something else. "If I can build a boat made entirely of materials that are fully recyclable materials and cross the ocean, why can't we build everyday household items that are cradle to cradle, rather than cradle to grave?" asks de Rothschild, grinning at the thought of turning trash to treasure.

Several companies are building equipment to transform beach trash into solid bricks and boards that can be used for building materials. Plastic bottles gathered during beach cleanups can be recycled for a variety of new uses. Getting at the small pieces inextricably mixed with plankton is much more challenging. Anything that could strain out the bits of plastic would take the plankton as well. "Better to stop the flow of plastic at the source," says Charles Moore, a retired businessman, surfer, and boat captain, one of the first to sound the alarm about the plasticization of the sea.

By chance, Moore encountered the Great Pacific Garbage Patch in 1997 while sailing to Hawaii. He was horrified when he came upon what appeared to be an enormous uncharted island, a horizon-to-horizon raft of floating flotsam. With his boat, the research vessel *Algalita,* as a seagoing laboratory, he started the Algalita Marine Research Foundation and has become a passionate ambassador for cleaning up the ocean, returning several times to explore and document thousands of square miles of the plastic potpourri. He is focusing on making people aware of the consequences of their use-it-once-and-throw-it-away habits, and inspiring actions that will prevent adding further insult to an already injured ocean.

On a 1999 expedition, Moore compared the number and weight of living plankton with the number and weight of plastic pieces in samples taken with fine-mesh nets pulled through the Garbage Patch. Life lost, six to one. For every pound of plankton, six pounds of trash! For every ton of living creatures, six tons of deadly junk!

Ten years later, at the 2009 TED (Technology, Entertainment, Design) Conference in Long Beach, California, Moore addressed a cheering audience wearing a hat and rainbow-colored necklace, both made from trash that he had relieved from the sea. Afterward, he showed me a jar full of recently sampled plankton, small animals lost in a blizzard of colorful flakes. "This just has to stop," he said. "We're killing the ocean."

As a biologist, I have explored many buckets of seawater, reveling in the micro-zoo that thrives in oceans everywhere—little jellies, larval fish, baby sea stars, jewel-like arrowworms, young crabs and shrimp—an endless parade of Mardi Gras characters, mostly translucent, often bioluminescent, always yielding something I've never seen before, and that maybe no one else has seen before, either. The sunlit surface of the sea is the place where most of Earth's oxygen is generated, where the great majority of carbon dioxide is extracted from the atmosphere, where food is generated to power the great ocean food webs. Does it matter that this critical part of Earth's life-support system is awash in plastic minestrone?

It definitely matters to the seabirds that mistake colorful bottle caps, cigarette lighters, and pieces of plastic foam, syringes, toy soldiers, and Lego parts for food. Scientists studying nesting albatross and other seabird colonies on Midway Island, part of the northwestern Hawaiian Islands, found thousands of dead chicks, their feather-fluffed corpses stuffed with hundreds of plastic bits. Ninety-five percent of fulmar carcasses washed ashore along the North Sea coast were stuffed with plastic, an average of 45 pieces per bird.

It also matters to the fish that snap up bits too small to be noticed by birds. Moore and his Algalita team examined the stomach contents of fish in the central Pacific and found that nearly all had at least some plastic, and one tiny lantern fish, 6.4 centimeters (2.5 inches) long, was packed with 84 individual pieces.

Even smaller pieces are engulfed by inch-long krill; ant-size copepods; and filter-feeding salps, clams, oysters, and mussels. Large plankton feeders such as whale sharks and manta rays swallow gallons of water at a time, plastic and all. Whether at the large, medium, small, or ultra-small scale, ingested plastic lumps, clumps, pellets, or microscopic mites kill by physically obstructing, choking, clogging, or otherwise stopping up the passage of food. That's bad enough, but there are other consequences.

A drum of unknown contents blights a Florida reef.

Plastic is made of petrochemicals that include various dyes and other additives, depending on their intended use. Although some may be inert and harmless, none were intended for consumption by living creatures. Plastics themselves may contain substances that mimic hormones and can become endocrine disrupters. There is some evidence that these "gender-benders" may be influencing the biological systems of marine organisms high on the food chain, including polar bears.

More troubling is the way that small pieces of plastic attract and concentrate toxins that are in the ocean—mercury, fire retardants, pesticides. Nurdles sampled in waters near Japan had levels of DDE (dichlorodiphenyldichloroethylene, a pesticide derivative), and PCB (polychlorinated biphenyl) a million times greater than were in the surrounding sea. Bioaccumulation of toxins occurs in food chains anyway, with the levels of mercury or other pollutants intensifying each time a small fish is eaten by a larger one. This effect is magnified enormously when toxin-enhanced plastic is consumed.

There appears to be no limit to how far down the food chain the accumulation of plastics and their toxic chemical baggage can go, reaching even the microbial swarms that dominate life in the sea. Plastics eventually break up into smaller and smaller pieces, but it appears that they are remarkably stable, retaining their identity and properties as plastic, even at the microscopic level. The consequences to ocean chemistry are simply unknown, but they need to be understood and factored in to the growing number of issues directly affecting the ocean's health, and thus our own.

Those who consume seafood should be asking another question: How far *up* the food chain do ingested plastics go? Does it matter if we eat oysters or anchovies or clams that have stuffed themselves with something other than little shrimp and algae? When toxin-loaded plastics are taken in by little fish that are consumed in large numbers by larger fish, and they in turn by larger ones still, it doesn't take long to deliver a concentrated dose of something that started out in the ocean in very dilute amounts.

THE ZERO WASTE INITIATIVE

In 2006, in response to growing awareness of the magnitude of the known and potential problems caused by marine debris, the U.S. Congress enacted the Marine Debris Research, Prevention, and Reduction Act. Two years later, a congressionally mandated report from the National Research Council summed up marine debris as a crisis, one that most likely will worsen through the 21st century. Its recommendation: zero waste discharge into the sea. The chairman of the committee that wrote the report, Alaskan scientist Keith Criddle, said, "We concluded that the United States must take the lead and coordinate with other coastal countries, as well as with local and state governments, to better manage marine debris and try to achieve zero discharge."

Andrea Crump, of the Marine Conservation Society in the U.K., has another angle. She points out, "Every single piece of rubbish has an owner and every single person can make a difference by making sure they take it with them."

It has taken us a while, but maybe the concept has finally sunk in. We can shift our trash, move it, cover it up, toss it into the sea, and turn our back, but everything connects. There is no "away" to throw to.

II. THE REALITY

THE OCEAN IS IN TROUBLE; THEREFORE, SO ARE WE

Jellyfish abound in Palau's saltwater lakes.

PREVIOUS PAGES: *Summer light illuminates Antarctic terrain.*

BIODIVERSITY LOSS: UNRAVELING THE FABRIC OF LIFE IN THE SEA

If the biota, in the course of aeons, has built something we like but do not understand, then who but a fool would discard seemingly useless parts? To keep every cog and wheel is the first precaution of intelligent tinkering.

—Aldo Leopold, *Round River,* 1953

Aliens looking for the nature of life on Earth would most likely start with the planet's dominant feature—the sea. "Follow the water," they might say. "All life requires water, and just look! The world is blue!" Diving in, they would quickly discover what Earthlings are at last coming to understand. The greatest abundance and diversity of life in the world—or as far as anyone knows, the universe—is out there, down there, from the surface to the greatest depths of the sea.

By itself, water is a combination of hydrogen and oxygen that, as a liquid, dissolves thousands of other compounds and picks up and transports enormous amounts of nonliving material, from salts to sand. But water alone does not generate oxygen, absorb carbon dioxide, or yield the simple sugars that are the basis of food production powering most of life on Earth. By itself, water does not produce the dimethyl sulfide molecules around which water gathers to form vapor that becomes clouds that in turn become rain, sleet, and snow. Microscopic photosynthetic organisms in the sea do all of these things and much more.

Curiously, much effort has gone into looking for life elsewhere in the solar system and beyond, always preceded by the question: Where is the water? With water, there could be life; no blue, no life. But learning about the nature of life in Earth's waters has been neglected over the ages, perhaps because humans are terrestrial, air-breathing creatures, complicated by the difficulty of accessing lakes, rivers, and the ocean below a few feet. Whatever the reasons, relatively little attention has been accorded to life in the realm of everything wet. That life includes the very small—the microbes—as well as most of the divisions of photosynthetic organisms and the bulk of the animal kingdom, the 30 or so phyla of creatures that do not share with us a characteristic vertebral column. Books about animals typically focus on our fellow vertebrates: mammals, birds, amphibians, reptiles, and sometimes a few fish. Largely neglected are the invertebrates, including the 500,000 or so kinds of insects and most of the creatures that occur only in the ocean—about half of the large divisions of animal diversity.

DOES BIODIVERSITY MATTER?

"Why care about the number and kind of creatures that live in the sea?

"What are all those creepy-crawlies good for, anyway, if you can't eat them or grind them up for oil or to feed to cows and pigs?"

"Who cares if some species go missing? The dinosaurs are gone, woolly mammoths are gone, there aren't any passenger pigeons any more, and the world keeps turning."

"What difference will it make if sharks or whales disappear? Or coral reefs, for that matter?"

I get lots of questions such as these as I go around the world expressing concern about how little we know of the nature of life in the sea, and how quickly, in my lifetime, losses have been building. We're losing not just individual species but entire ecosystems packed with creatures we have not yet named and whose place in the greater scheme of things we don't yet know. The best way to answer is to ask in return, "Have you thought about moving to Mars? No pesky creepy-crawlies there! Do you like to breathe? Are you ready to give up water? A congenial climate? A prosperous future for yourself and people you care about?"

These things and many more depend on the mostly blue, biologically rich world we have received. We are newcomers on a planet that has changed many times over the ages. The matrix of living things that shape the world has changed shape repeatedly over hundreds of millions of years in response to occasional planetary upheavals as stupendous as the meteor strike that eliminated dinosaurs and opened the way for primate prosperity. Recovery and regrowth from these upheavals takes time and requires diverse forms of life in order to rebuild—and attain stability.

The smaller the gene pool, the greater the vulnerability of life to disease, storms, changes in climate, or other natural ups and downs. The greater the variety, the more likely it is that somewhere in the mix there will be survivors that can cope and prosper. This applies even with individual species. If all buttercups bloomed at the same time, none would be there for bees to pollinate if bee populations were late in arriving. If all humans were equally susceptible to bubonic plague or the measles or tuberculosis, all of us might have perished by now.

Harvard biologist E. O. Wilson gives his take on why biodiversity matters in his book *The Diversity of Life:*

> Biological diversity—"biodiversity" in the new parlance—is the key to the maintenance of the world as we know it. Life

Slender branches of a coral shelter dozens of other species.

in a local site struck down by a passing storm springs back quickly because enough diversity still exists. Opportunistic species evolved for just such an occasion rush in to fill the spaces. They entrain the succession that circles back to something resembling the original state of the environment.

This is the assembly of life that took a billion years to evolve. It . . . created the world that created us. It holds the world steady.

The way that humankind is driving the loss of biodiversity on land is well documented—and cause for serious alarm. Every bird, mammal, insect, or tree, every kind of wild potato or rice or corn now missing from the global treasury since our arrival on the planet translates to less stability, fewer opportunities to find solutions to questions relating to food security, health, and the overall well-being of the planet.

As a child, I enjoyed disassembling things—toys, clocks, an old pump—and can still hear my father saying, "Did you save all the pieces? Do you know how to put that back together again? Can you make it work?"

"Who but a fool," asks Aldo Leopold, "*would discard seemingly useless parts?*"

ASSESSING THE DAMAGE

Concerns about species loss have not been limited to scientists and conservationists. At the World Summit on Sustainable Development in Rio de Janeiro, Brazil, in 1992, a treaty—the Convention on Biological Diversity—was produced that by 2007 had been ratified by 190 nations. The goal of the treaty is simply to support efforts to sustain the diversity of life on Earth. Data assembled by thousands of scientists and hundreds of institutions working through the International Union for Conservation of Nature (IUCN) provide insight into how much has gone missing in recent years owing to human impacts and help gauge the level of ongoing threats.

So far, only about 50,000 species out of millions have been assessed, with priority given—not surprisingly—to flowering plants and terrestrial vertebrates. Nonetheless, the trend is clear. One in every eight birds is in danger, one in every four mammals, and one in every three amphibians. Up to 73 percent of all flowering plants are at risk. Wilson predicts that at the present rate, fully half of the planet's plant and animal species will be gone by the end of this century, noting that "the loss of genetic and species diversity . . . is the folly our descendants are least likely to forgive us."

At the start of the 20th century, some parts of the ocean's diversity had already been discarded, notably the Steller's sea cow, the Atlantic gray whale, and the great auk. Many species had declined drastically, and coastal development and destructive fishing had no doubt eliminated innumerable small organisms whose passing went unnoticed and unsung.

According to a 2008 IUCN Red List report on the status of the world's marine species, about 17 percent of sharks and their relatives now are threatened and an additional 13 percent are nearly threatened, with too little known about an additional 47 percent to be able to provide a meaningful assessment. Ironically, most of the information obtained for the research came from the very fishing activities that are the primary threat to the sharks' survival.

Among the many marine species in peril are all groupers, devastated by overfishing. Some, including a once common fish that I knew as a child in the Florida Keys, the Nassau grouper, are considered endangered and are now fully protected in U.S. waters. During a 1998 research expedition in the underwater laboratory *Aquarius* near Key Largo, Florida, I was excited to find one of these increasingly rare animals. The youngster was as curious and attentive as those I had encountered many years before. Unfortunately, he was the only one seen in the area by six of us diving day and night for a week.

Mammals, too, are vanishing. There are only about a hundred of Mexico's beloved little porpoises, the vaquita, living in the upper reaches of the Gulf of California; a subspecies of the Hector's dolphin in New Zealand has about the same number remaining. Similar numbers account for the western Pacific gray whales. About 300 northern right whales harbor the only hope for their kind, forever.

All species of sea horses are threatened for two major reasons: habitat destruction, mostly by coastal development and shrimp trawling, and deliberate capture to serve a high-end Asian market for dried sea horses—ground and used for various medicinal applications, including as an aphrodisiac.

Marine reptiles are also affected. For instance, six of the seven sea turtle species are threatened, and all are at historic low levels of population. In half a century, 97 percent of leatherback sea turtles have been eliminated in the Pacific Ocean, their primary home.

In 2007, corals were added to the IUCN Red List of threatened species for the first time—not because these colonial organisms were not threatened earlier, but because there had been a common misperception that marine species are not as vulnerable to extinction as terrestrial ones. A study of corals and algae in the Galápagos Islands showed otherwise, providing convincing evidence that three species of coral and ten kinds of macroalgae that are unique to the islands could soon disappear forever.

In the early days of the IUCN Red List, only marine mammals were considered vulnerable among marine organisms. That has changed with awareness that the ocean, even more than the land, has been altered by human activity, but the changes have largely gone unnoticed because people can't readily see the damage. Moreover, records of what the ocean was like before serious exploitation are sketchy. Jane Smart, head of the IUCN Species Program, noted in a recent report that marine ecosystems are vulnerable to threats at all scales: globally through climate change, regionally from El Niño events, and locally when overfishing removes key ecosystem building blocks.

HOW MANY MARINE SPECIES ARE THERE?

To understand the magnitude of the problem, it helps to know how many marine species exist in the first place. How many kinds of organisms are we talking about?

In the first century A.D., Roman naturalist and historian Pliny the Elder thought he knew. With a list of 176 species of fish and other sea creatures, he was certain that "in the ocean . . . nothing exists which is unknown for us."

Seventeen centuries later, the number of plants and animals discovered in the sea had grown to thousands, but in 1858, British scientist Edward Forbes expressed the view held by many that whatever lived in the ocean was limited to the region above 300 fathoms, or 550 meters (1,800 feet). He thought nothing alive could cope with the high pressure, lack of sunlight, and numbing cold in the vast realm that he called the azoic zone. Some scientists believed that below a certain depth, increasing pressure made the ocean impenetrable to things falling from the surface, so much so that sunken ships, the bodies of men lost at sea, and treasure of all sorts endlessly drifted without ever reaching bottom.

That myth and the idea of a "lifeless zone" were solidly put to rest during the four-year British *Challenger* expedition (1872–76), the first global scientific exploration of the sea. Instructions supplied to the captain sound like the launching of a *Star Trek* adventure: "You have been abundantly supplied with all the instruments and apparatus which modern science and practical experience have been able to suggest and devise . . . you have a wide field and virgin ground before you." With a long list of unanswered questions—"How deep is the sea? What is it like on the ocean floor? Is there life at great depths?"—the scientists had a simple but comprehensive mandate: literally, to boldly go where none had gone before, "to explore all aspects of the deep-sea."

H.M.S. *Challenger,* a motor-assisted sailing vessel, covered 127,584 kilometers (68,890 nautical miles) in all of the major ocean basins except the Arctic, charted 362 million square kilometers (140 million square miles) of ocean floor, and gathered new data on ocean currents, temperatures, and sediments in depths to nearly eight kilometers (five miles). Using nets, dredges, hooks, and other remote-sampling devices, the crew pulled in many thousands of species, including 4,417 kinds of previously unknown plants and animals.

Since the *Challenger* expedition, thousands of scuba divers and hundreds of submarines and remotely operated vehicles have sampled places in the ocean not reached until 20th-century technologies made access possible. These explorers have greatly increased our knowledge of how many kinds of life, and what kinds of life, exist in the ocean. And yet this research is just a starting point. By 2000, about 1.5 million kinds of organisms had been described and named as species living on Earth. Of these, 25,000 or so were fish, both freshwater and marine. The number of marine species identified so far, from blue whales and giant tunas to shrimp, microscopic plankton, and even bacteria, totals about 250,000.

At first glance, it doesn't make sense. How can there possibly be greater diversity of life on the land than in the sea? Life requires water, and most of the water on Earth is ocean. About 97 percent of the planet's living space extends three-dimensionally from the surface of the sea to the deepest cracks, canyons, and trenches, almost 11 kilometers (7 miles) down. Life originated in the sea. It has had four billion years or so to develop every conceivable variation on the theme of life, plus many more that we have not yet been able to imagine.

The fact is, we have barely begun to find and name the vast numbers of species in the ocean, much less understand how they work together. And while the focus on species is a good place to start, it is a little misleading, since "biodiversity" goes beyond the splintery ends of life that we call species. Peter H. Raven, chairman of a national committee overseeing a survey of the biodiversity of the United States, defines biodiversity as:

> . . . the sum total of all the plants, animals, fungi and micro-organisms in the world, or in a particular area; all of their individual variation; and all of the interactions between them. It is the set of living organisms that make up the fabric of the planet Earth and allow it to function as it does, by capturing energy from the sun and using it to drive all of life's processes; by forming communities of organisms that have, through several billion years of life's history on Earth, altered the nature of the atmosphere, the soil and the water of our planet; and by making possible the sustainability of our planet through their life activities now.

ORGANIZING LIFE BY ITS GENES

Species—actually, the genetic material that makes each definable—are the fundamental building blocks of diversity, a natural level of organization. The biological definition of a species is "a population of organisms whose members are able to interbreed freely under natural circumstances." (There are plenty of unnatural circumstances in which artificially manipulated species have developed into something far different from the original stock, from corn, rice, and beans to cats, dogs, chickens, and cultivated salmon.) Individuals within a species can look very different from one another but still interbreed, and therefore can still be considered members of the same species.

Insights developed in the 20th century concerning the genetic material that forms the basis of life revolutionized understanding of the relationships among organisms, notably among many small organisms, such as nematode worms and all microbes, that for the most part look very much alike, although they may be genetically very different.

A group of closely related species is referred to as a genus, and the scientific name of most creatures includes both genus

Sea stars adorn a Galápagos reef.

and species. The closely related bull sharks and gray reef sharks are both in the genus *Carcharhinus,* and each has its own name: bull sharks are *C. leucas,* gray reefs are *C. amblyrhynchos.* Closely related genera are grouped together in families; groups of families are called orders. Related orders are grouped as classes, and one or more class is called a phylum for animals, a phylum or division for plants—the highest level of organization below the grand category called kingdom.

To *Challenger* scientists, it was simple. There were two kingdoms, one for plants, one for animals. Today, our increasing knowledge of the diversity of life means seven kingdoms are recognized, grouped into three great domains, all of them abundantly represented in the sea:

Domain Archaea

These recently recognized, exquisitely small organisms resemble bacteria—microscopic, with no organized nuclei within their cells—but they have a significantly different genetic makeup.

Kingdom Archaeabacteria. First described from hot springs in Yellowstone National Park, these microorganisms abound in deep-ocean hydrothermal vents, petroleum deposits far underground, and the intestines of cows, among other places.

Domain Monera

These single-celled organisms have no organized nuclei and look like archaea but are genetically distinct.

Kingdom Monera. Grouped here are at least a dozen major kinds of bacteria, including those once known as blue-green algae. Only about 4,000 Monera had species names in 1990, and what constitutes a species of bacteria and other microbes is still being debated, but it is estimated that there are a mind-boggling 10^{30} distinctive forms of Monera. Most occur in small numbers in a dormant form, and most are in the ocean. In samples taken by biologist Craig Venter in the transparent waters of the Sargasso Sea in 2004, 1,800 species of microbes and over 1.2 million new genes were discovered. Every spoonful of ocean sampled turned up millions of individual microbes and a host of new species.

Domain Eukaryota

This is where humans come in, along with all other organisms that have within their cell or cells a membrane-bound nucleus containing genetic material organized into chromosomes.

Kingdom Protista. To this group of organisms, coupled with blue-green bacteria in the Kingdom Monera, and certain members of the Kingdom Chromista—the coccolithophorids—we owe Earth's congenial, oxygen-rich atmosphere. Most phytoplanktonic organisms are represented here, as well as red and green algae and numerous single-celled animals.

Kingdom Chromista. Sometimes lumped with plants, this large assemblage of mostly marine organisms includes microscopic diatoms and giant kelps that may grow to be 100 meters (330 feet) long. They characteristically have golden pigments and chlorophyll c, and they do not store energy as starch—a key characteristic of plants.

Kingdom Fungi. Only a few hundred of the 100,000 or so kinds of fungi known have been identified in the ocean, but many more are believed to exist. None look much like mushrooms or bread mold.

Kingdom Plantae. About a dozen phyla within the Eukaryota are defined as "plants"—multicellular photosynthetic organisms with cellulose in their cell walls and a life history that typically alternates a generation with a single set of chromosomes (haploid) with another that has chromosomes joined (diploid). Pine pollen is haploid; pine trees are diploid. About 250,000 kinds of mosses, ferns, trees, flowers, and other plants occupy the land; mangroves, salt marsh shrubs and grasses, and about 60 or so truly marine flowering plants lumped as "sea grasses" live in the sea.

Kingdom Animalia (Metazoa). Technically, animals are multicellular, nonphotosynthetic organisms that lack cell walls and have genetic material within nuclei. Most kinds form tissues with specialized organs and have embryos with two sets of chromosomes. Debates are ongoing about how to organize and group the thousands of creatures included in this kingdom, ranging (in the sea) from gelatinous jellyfish and comb jellies to sea stars, crabs, clams, fish, whales, and scuba divers.

All domains and kingdoms and most phyla have at least some representation in the sea. Only about half of the phyla of animals have terrestrial or freshwater species, but nearly all of the 36 or so known are found in the sea.

So how many species really live in the sea? Is there a way to determine how many lived there in times past? Might it be possible to predict the future diversity and abundance of life in the sea?

THE CENSUS OF MARINE LIFE

In 2000, a group of scientists set out to answer these and other questions in an ambitious endeavor known as the Census of Marine Life. With initial funding from the Alfred P. Sloan Foundation, a network of more than 2,000 researchers in 80 countries developed a ten-year plan to assess and explain the diversity, distribution, and abundance of life in the sea.

Hundreds of institutions and agencies have helped support the effort. The endeavor includes mining existing records in museums and private collections as well as gathering new data through expeditions exploring habitats and groups of species throughout the world, from polar seas to tropical reefs. A major database, including more than 15 million records, provides the framework for the Ocean Biogeographic Information System (OBIS), where information about what lives where in the oceans is available electronically.

The History of Marine Animal Populations (HMAP) project has involved constructing a record of marine life during the past 500 years, starting at about the time that human activity began to have a major impact on life in the sea. Understanding the way things were provides a vital measure of change, especially important when evaluating the unprecedented impact of industrial-scale fishing in the 20th century.

The Future of Marine Animal Populations (FMAP) study takes data from a wide number of sources to analyze and predict future trends.

Sponges such as this colony in Belize filter the ocean for food.

The main focus of the census is to address the present circumstances in six major ocean realms. There are many ways to divide up the blue part of the planet, but it is useful to include here the approach adopted by the census:

1. *Human Edges.* Generally running from the high-tide line to the bottom of the continental shelf, this is the area that is embraced within the Exclusive Economic Zones of most countries. The region from the high-tide line to the 10-meter (33-foot) depth is considered the nearshore zone. It includes about one million kilometers (620,000 miles) around all of the coastlines across all latitudes and climates, from coral reefs and sea grass meadows to

narrow icy shelves. The coastal zone continues seaward from the nearshore to the edge of the shelf.

2. *Hidden Boundaries.* This term mostly encompasses the little-known slope region of the continental margins extending outward to ocean basins, including abyssal plains.
3. *Central Waters.* The largest habitat on Earth, the open sea is home to at least 40 percent of the planet's primary productivity biomass. The illuminated light zone extends from the surface to a depth of 200 meters (660 feet) and the dark zone from there to the ocean floor. The terminology applies primarily to the penetration of sunlight, as bioluminescence expert Edith Widder made clear: "The deep sea is often described as 'a world of eternal darkness.' That is a lie. While it is true that sunlight does not penetrate below 1000 meters, . . . there are lots of lights—billions and billions of them." This is the realm of clear jellies, salps, swimming snails, and arrowworms—a glass menagerie of life that has no parallel on the land. It is home not only to some of the smallest creatures on the planet but also to the longest, a translucent siphonophore called *Praya* that may extend as much as 40 meters (130 feet).
4. *Active Geology.* This category includes seamounts, hydrothermal vents, and cold seeps, with life-forms ranging from chemosynthetic bacteria to the ecology of the numerous and widespread undersea mountains. The Census of Seamounts considers the role of these great undersea islands in developing unique species as well as in serving as way stations for undersea species.
5. *Ice Oceans.* Ocean ice in Antarctica surrounds the land, while in the Arctic, ocean ice is surrounded *by* land, yet there are many similarities in terms of the challenges faced by organisms living at opposite ends of the Earth.

6. *Microbes.* Every drop of ocean water contains microbes—about 20,000 different types in a single liter of water, according to recent census data. They represent not only the oldest forms of life on Earth, but also the category with the capacity to change most rapidly, with many generations coming and going in a matter of days, even hours.

THE UNSEEN MILLIONS

But the question remains: How many ocean species are there?

Not only do we not know, but the answer may be unknowable for the foreseeable future, given the size of the challenge, especially at the microbial level. Even with the ambitious effort of the Census of Marine Life and the combined institutional efforts of various navies, oceanographic research institutions, and individuals, currently less than 5 percent of the ocean has been seen, let alone explored; for the deep sea, below 305 meters (1,000 feet), the figure drops to about one percent.

The magnitude of what remains to be discovered is exemplified by the discoveries of Richard Pyle, a truly intrepid biologist who uses cutting-edge re-breather diving systems and special mixes of gases to explore the almost-light/almost-dark region of the sea known as the twilight zone, from 100 to 200 meters (330 to 660 feet) down. He finds new species of fish at the rate of 12 to 13 (and sometimes as many as 30) per hour. For invertebrates, the number is at least ten times that.

Given the number of new forms of life discovered in regions examined for the first time, it is now estimated that the ocean holds at least ten million species. Some believe the number may be closer to a hundred million—not counting the microbes that eclipse all other forms of life in terms of numbers and sheer mass.

So with all those species, some might ask the old question, "So who cares if we lose a few species? There appear to be plenty out there that we don't even know about."

While there are clearly many more unknown than known species in the sea, we may be losing them faster than we are discovering them. Already about half of the coral reefs worldwide have either disappeared or are in a state of sharp decline, and as they disappear, unique wedges of genetic diversity slip away as well. In the deep sea, ancient forests of coral, sponges, and their many associates are being destroyed as fishing boats drag trawls across the ocean floor to catch fish as old as the fishermen's great-grandparents. Without understanding the living machinery that keeps us alive, and rather than doing our utmost to keep that machinery running smoothly, we probably could not have dreamed up more effective ways to disrupt it than we have already, albeit unwittingly.

Knowing the value of the diversity of ocean wildlife may be the key to saving it. In a 2006 report in *Science,* 14 scientists assessed the impacts of biodiversity loss on the "free" services ocean ecosystems provide to us. Biodiversity loss was shown to impair the ocean's capacity to provide food, maintain water quality, and recover from perturbations. On the positive side, in localized areas where long-term data were available, restoration of biodiversity yielded a fourfold increase in productivity and decreased variability by 21 percent.

Rather than ask who cares if we lose this species or that, we might ask what difference it would make if humankind were to disappear, In *The World Without Us,* Alan Weisman speculates about the hauntingly swift transformation of our cities and farms into jungles of recovering wilderness. Without humans extracting millions of tons of wildlife from the sea or putting millions of tons of garbage in, many depleted species would most likely rebound within a few hundred years. Some might respond much more quickly, as has been the case in marine protected areas where fishing is restricted. Appreciable recovery of more and larger fish and overall increased diversity in protected areas can happen within a scant two years.

The bottom line answer to the question about why biodiversity matters is fairly simple: The rest of the living world can get along without us, but we can't get along without them. Diminishing the diversity of life as we are now doing translates to diminished chances for our continued prosperity. John C. Sawhill, president of the Nature Conservancy from 1990 to 2000, gave as good a reason as any for not losing more of the splendor of life: "In the end, our society will be defined not only by what we create, but by what we refuse to destroy."

Kuwait, 1991: The essence of ancient forests up in smoke.

DRILLING, MINING, SHIPPING, SPILLING

6

What is the use of knowing how deep [the sea] is unless we know what is at the bottom of it? . . . Where was the mechanical skill that would contrive for us the means of bringing from miles below . . . the feathers from old ocean's bed, be it ooze, or mud, or rock, or sand . . .

—Matthew Fontaine Maury, *The Physical Geography of the Sea,* 1855

It is a staggering thought, keeping a ship longer than two end-to-end football fields steady enough on a rolling ocean to allow more than five kilometers (three miles) of rigid pipe, 30 to 41 centimeters (12 to 16 inches) in diameter, to be deployed like a very long straw to the seafloor. More astonishing is the idea that once in position, the system could somehow slurp up enough rocks to make a deep-sea mining operation pay off. With a small group of scientists and engineers touring the massive ship *Glomar Explorer,* in Long Beach, California, that is the story I heard in 1974, close to the time that the real mission was revealed—the salvage of the Soviet submarine, *K-19,* sunk in April 1968 with nuclear torpedoes and missiles on board. To avoid interference with the Central Intelligence Agency's covert salvage operation, code-named Project Jennifer, an elaborate cover story had been concocted about aviation pioneer Howard Hughes leading an effort to mine manganese nodules.

Holding out a double fistful of lumpy black rocks, our guide smiled convincingly when he said, "Manganese nodules such as these are rich in nickel, cobalt, copper, even silver, gold, and platinum, as well as iron and manganese. Thousands of square miles of ocean floor are littered with these treasures, just waiting for the right technology to be able to go get them. Now we can do just that." He showed us a film of the crew deploying the drilling string through the "moon pool," an improbable-looking opening into the ocean right through the midsection of the ship. We were ushered into the computer room, a spacious place with aisles of slots holding data cards that provided three redundant modes to control the ship's position, with a fourth critically important "manual override." Should the computers fail to hold the ship steady, a human being could take over. We were *not* taken to see the giant three-million-kilogram (six-million-pound) manipulator "claw"—effective for salvaging sections of a submarine but hardly the kind of instrument needed to pick up potato-size rocks.

Manganese nodules have been known to be abundant in deep water globally since their discovery during the 1872–76 voyage of the H.M.S. *Challenger,* but they attracted little attention for a century because of the expense and difficulty of retrieving them. Shortages of valuable metals and greatly improved technologies for accessing the deep sea provoked a number of countries and companies to begin to take deep-sea mining seriously in the 1970s. That reality made the fantasy mission of the *Glomar Explorer* seductively convincing, and the technologies developed for this and other military operations during World War II and the Cold War that followed provided a boost for those truly interested in deep-sea mining and drilling, for fishermen, for scientists—for all who had interests in, on, and under the ocean.

Hundreds of millions of dollars were invested in technologies to realize the dream of mining manganese nodules in the 1970s and early 1980s. The vision of great economic rewards was so compelling that the Law of the Sea negotiations in the 1980s were effectively stalled as nations haggled over who could claim what on the deep-sea bed. The United States and other nations with the leading technological capabilities to access the deep sea clearly were not eager to give up their edge to benefit countries without such advantages, while other nations were not pleased about the prospect of yielding their claim to the ocean's wealth, even though they could not get there.

To me, it seemed more than a little premature to carve up and exploit vast areas of the planet that had not yet been mapped or seen, with essentially no understanding of the importance of anything that might be there other than marketable minerals. An engineer assigned to evaluate prospective mining sites confided to me that nothing would be harmed by using the large bulldozer-like machine he was helping to design to gather up nodules, as there was "nothing alive down there except for a few sea cucumbers and brittle starfish that nobody cares about." Now we know that the diversity of small creatures living in the sand and mud of the deep sea can exceed the richest places known on the land.

I was astonished in 1980 listening to the U.S. representative to the Law of the Sea deliberations tell a Washington, D.C., audience that the impact of deep-sea mining would be negligible, as "there is nothing on the bottom of the deep sea but green slime."

As a biologist keenly aware of the amazing variety and abundance of life everywhere anyone had looked in the ocean, I was mystified at the notion that the deep sea is covered in "slime," let alone *green* slime!

Rather than shrug off impacts of what would surely be destructive mining activities in the unexplored regions beyond where nations claimed jurisdiction, I really like the model provided by the Antarctic Treaty. That vast, white wilderness, all of the land and ice shelves south of 60°S latitude, has been respected internationally since 1961 as a place where scientific investigation is allowed but military activity is excluded. A Convention on the Regulation of Antarctic Mineral Resource Activities was signed in 1988 but subsequently rejected. In 1991, a new Protocol on Environmental Protection to the Antarctic Treaty was approved, prohibiting all but scientific activities related to mineral resources. Why not similar "precautionary principles" for the deep-sea bed, I have long wondered—but wished in vain.

While some protective policies have been built into the Law of the Sea Treaty, the allure of "silver, gold, copper, nickel, cobalt" waiting to be scooped from the seabed provoked development of specific language for deep-sea mining. The treaty established an International Seabed Authority to authorize exploration and mining and to collect and distribute mining royalties. As of mid-2009, 158 nations have signed and ratified the treaty, but the United States is among 21 that have signed but withheld ratification, largely owing to lingering concerns about giving up more than might be gained.

STRIPPING THE DEEP

During an eight-year project in the 1970s, an international consortium collected tons of manganese nodules from the abyssal plains of the eastern Pacific and succeeded in extracting significant quantities of nickel, copper, and cobalt, but not enough to justify scaling up to full commercial operations. Since the mid-1980s, interest in mining manganese nodules has diminished, but in recent years, enthusiasm for the recovery of polymetallic crusts associated with volcanic activity around hydrothermal vents in the deep sea has grown.

Under the Law of the Sea, nations may claim jurisdiction over an Exclusive Economic Zone of the ocean, a region extending seaward 200 nautical miles from the edge of their territorial sea, 12 nautical miles from shore. Foreign nations have the freedom of navigation and overflight, subject to regulation by the coastal states, but in general, the states have sole exploitation rights over the natural resources within the region. Rather than dealing with the known restrictions, expense, and continuing uncertainties of mining in international waters, attention is shifting to exploiting areas under national jurisdiction, where regulations are better defined and, in some countries, more relaxed.

A Canadian company, Nautilus Minerals, is focusing on a region off the coast of Papua New Guinea where active hydrothermal vents have formed crusts rich in several metals including the old favorites—gold and copper. The company has leased areas where large machines will be used to strip-mine within the top 20 meters (66 feet) or so of the ocean floor, then use a hydraulic pump system to raise materials to the surface.

The endeavor is enormously challenging at every stage, from exploration survey and collection, to movement of the minerals from the seafloor to the surface, to final transport to the processing site on land or at sea. Processing is another major undertaking, with serious environmental issues wherever it takes place. Thereafter, there are steps in storing and eventually shipping the metals for sale, complemented by the business of marketing them.

Applauded as a monumental technical achievement, there are concerns about the consequences of mining polymetallic crusts because of the inevitable disruption to the seabed. Every area of the deep sea thus far explored has yielded surprising discoveries, especially concerning the role of microbes and other marine life in basic ocean chemistry. The effects of large-scale mining, or even of small scale in sensitive areas, are simply unknown. Does not knowing the consequences mean there is nothing to worry about?

Impact assessments have been made at the targeted mining sites in Papua New Guinea, and there are proposals to deliberately leave some of the leased areas intact, but even with these provisions, there will be unavoidable loss of marine life through mining and the release of toxic cuttings and fine sediment into the water column. Jochen Halfar, a geologist from the University of Texas, noted in a 2007 report from *ScienceDaily*, "We need to act now to establish scientific and legal methods to protect these sensitive ecosystems and minimize the potential environmental impact of this industry. . . . the prospects for regulation of underwater mining are not good."

Ultimately, the decisions about mining the deep sea will most likely be based on the usual short-term perception of economic values. Rain forests sized up only for their lumber or for the underlying land that can yield what are looked upon as "higher returns" when planted with soybeans or pasture for cows are more likely to be clear-cut than those where the economic significance of the diverse, living forest is taken into account, layered on top of the aesthetic and ethical importance of protecting species and ecosystems. Similarly, unless there are champions who can articulate—and decision-makers who can understand—the literally "priceless" importance of the intact living systems in the deep sea, the value of rocks will almost certainly trump whatever might be lost.

Although plans for mining the deep sea continue to evolve, sand mined from beaches, dunes, and shallow nearshore areas is used in a growing number of businesses—as fill to "develop" coastal shoreline, as beach "nourishment," as road and building material, even as part of glass production and, in some cases, the extraction of minerals mixed with whatever else makes up what some just call dirt. Problems abound in the put-and-take relocation and

Since the '50s, platforms have extracted oil and gas from the sea.

consumption of sand, including the loss of homes for many millions of small creatures that live among the grains, destabilizing coastal areas where sand is removed, and—just as in mining the deep sea—stirring sediments that pollute the surrounding water.

DRILLING FOR ANSWERS

Drilling technologies developed for commercial and military applications have greatly benefited the search for knowledge, and in some measure, vice versa. In 1964, the National Science Foundation funded a project directed by the Joint Oceanographic Institutions for Deep Earth Sampling (JOIDES), a consortium of U.S. oceanographic institutions. Over the next several years, about 600 cores were drilled throughout major ocean basins, with stunning results. Analysis of the long, cylindrical core samples confirmed that the present ocean basins are relatively young, and demonstrated that the Mediterranean Sea completely dried up between 5 million and 12 million years ago. They established that Antarctica has been ice-covered for at least the past 20 million years and that the Arctic ice cap was significantly larger 5 million years ago.

These are big answers to big questions, profoundly important in understanding the nature of the planet and gaining perspective on our place in Earth's deep history. So significant were the discoveries that a new program got under way in 1984, the Ocean Drilling Program, supported by a consortium of 21 nations working from the drill ship *JOIDES Resolution.* The ship recently was transformed by a $100 million investment into a unique floating laboratory operated as a part of the Integrated Ocean Drilling Program with principal support from the U.S. National Science Foundation and Japan's Ministry of Education, Culture, Sports, Science, and Technology.

In March 2009, the ship sailed from Hawaii on a two-month voyage with 30 scientists from six nations, 25 technicians, and 66 crew from many more nations for the first of two expeditions planned for the year. The goal? To drill down through thousands of feet of ocean to extract cylindrical plugs out of the ocean floor and decipher the record of the past in deep-sea sediments. It requires the best detective work ever devised, a forensic analysis of what life was like from the present through the past 55 *million* years.

The fossils embedded within the mud and rock recovered in the core samples can be read to understand the nature of the climate in times past. Co-chief scientist Heiko Pälike likens the ocean-drilling operation to the space program in terms of its importance to understanding Earth. In a sense, the *JOIDES Resolution* is a time machine, with scientists leading the way on a journey into Earth's deep history.

"A thousand years" rolls off the tongue easily, as if we might actually get our minds around life as it was a *thousand years ago* when there were fewer than 300,000 people on Earth, and Viking longboats were prowling the coasts of northern Europe and North America—about the time that gunpowder was invented in China . . .

Ten thousand years is harder still to truly grasp. Imagine the magnitude of change in a hundred centuries!

Or how about a *thousand* times a hundred years? Or 200,000 years ago, about the time that modern humans are thought to have appeared.

Astronomer Carl Sagan challenged audiences to imagine compressing all of the 14 billion years since the birth of the universe into a single year. Earth began four and a half billion years ago, sometime in September. The first forms of life appeared late in the month and remained entirely microbial until far into November; significant oxygen began to develop in the atmosphere on December 1. The first fish swam in ancient seas on December 19; dinosaurs arrived on December 24 and became extinct on the 29th. During this era, representing more than 100 million years, sunlight fixed and stored by photosynthetic organisms on the land and in the sea formed the basis for much of the coal, oil, and gas now being tapped for fuel. The depth of the core samples being taken in 2009 from the *JOIDES Resolution* will reach back to sometime on December 29. On December 31 at 11:59, cave paintings were made in European caves. All of civilization as we know it occurred during the last minute of the last day of the last month of the year.

In 50 real years, the thinnest imaginable slice of Sagan time, our species has unwittingly managed to nudge the way the world works, with outcomes yet unknown. But it is looking good for the microbes.

FUELING CIVILIZATION

Of all the extractive industries now targeting nonliving ocean "resources," none have provoked more attention, positive and negative, than those involving oil and gas. For all but the latest tiny sliver of human history, the existence of naturally occurring petrochemicals has had a vanishingly small impact on civilization. Since the 1950s, however, oil has become the world's most important energy source, displacing its next of kin, coal, largely coincident with the growing importance of petroleum-based transportation, military and domestic. The high energy yield, relative ease of transport, and abundance of oil boosted global economies and provided the underpinnings of prosperity on many fronts. Petroleum products slipped into our everyday lives as the raw materials for plastics, pharmaceuticals, solvents, fertilizers, pesticides, cosmetics, and much more.

Any of us who fly airplanes, buy groceries in a supermarket, use plastics in any form, sprinkle commercial fertilizers on lawns or plants that we grow for food, or do almost anything that requires moving goods almost anywhere around the world are beneficiaries of the industries many love to hate—oil and gas—from the "upstream" exploration, development, and production of crude substances fresh from the ground or sea to the "downstream" side involving oil tankers and refiners, as well as all that in-between activity—pipelines, service, and supply.

There are many explanations of how in one century, fewer than two billion people more than tripled in number; how for the first time, flight became possible, not only across the land and sea but to the moon, Mars, and beyond; how gaining access to the deepest ocean, even 4,300 meters (14,000 feet) down under North Pole ice, became possible; how food production jumped to unprecedented levels; and communication became easy not just with next-door neighbors but with new neighbors living on the other side of the planet. Whatever the explanations, these and many other aspects of the 20th and 21st centuries would not—could not—exist but for the fossilized solar energy that we have burned to power our way from where we were to where we are now.

Civilization currently thrives on oil-based economies and is continuing to do so, despite herculean efforts to move away from fuels that pollute the planet today and potentially will shorten the number of tomorrows our species will have.

In a geological instant, people now alive have managed to consume millions of years' worth of fossil forests and trillions upon trillions of microbes that prospered in ancient oceans. We have burned through much of the accumulated carbon reserves of deep time to provide energy for a few generations of humans. Knowing this, and knowing that there are only so many compressed, multimillion-year-old forests and fossilized, liquidized microbial masses sequestered in Earth's treasury should win the rapt attention of even the most complacent couch potato. As it is, demand for petroleum products continues to rise, while the search for effective alternatives goes on.

Use of oil and gas products runs deep into human history but did not begin to expand seriously until early in the 20th century. As gasoline-powered vehicles became popular, so did the need for fuel to run them. Previous sources of energy—wood, coal, whale oil—probably would not have done the job.

By the end of World War II, the oil industry was firmly established in the United States and globally, but it did not begin to move into the sea until 1947. That year, the Oklahoma-based company Kerr-McGee completed the first drilling rig out of sight of land, in the Gulf of Mexico. Initially, land-based equipment was adapted for offshore use, but it did not take long for systems to be designed specifically for use in the ocean, and the Gulf of Mexico became the great laboratory for developing expertise for offshore operations.

In the 1978 book *Commercial Oil-Field Diving,* author Nicholas Zinkowski wrote:

> The diving business has grown as swiftly and as dramatically as the offshore oil industry, which is almost solely responsible for the tremendous surge in diving activity and technology. There is virtually no phase of the offshore oil industry—exploration, drilling, pipelining—that does not require the services of dependable professional divers. . . . The advances to date . . . indicate that a 1,500 foot-operational depth is at hand, but the physiological risks faced by divers in extremely deep water have yet to be fully evaluated. The financial costs of putting a team of divers down in 1,500 feet of water . . . for just a few hours of work . . . could easily approach one half a million dollars.

Even so, drilling operations were already exceeding those precarious diving limits. Near the end of the book, a chapter is devoted to "Articulated armor, submarines, manipulator bells and remote controlled vehicles," and features a photograph of the diving system called Jim, named for Jim Jarrett, the British diver who first used the prototype version in the 1930s. Made of metal with articulated arms and legs, Jim resembles an overstuffed version of an astronaut suit, and like systems used in space, has air delivered from a re-breathing system carried like a backpack.

As a seagoing scientist, I'm drawn to the idea of being able to pop down to a thousand feet or so without worrying about decompression. So, in 1979, on a book assignment with the National Geographic Society, my co-author Al Giddings and I talked Oceaneering International, owners of the Jim suits, into letting me try one. No one had used such a system for scientific exploration before.

I had already spent thousands of hours underwater using scuba, re-breathers, diving helmets, submarines, even living for weeks at a time in several underwater laboratories. But using Jim was a totally different kind of experience. Although designed for salvage operations and adapted for oil field work, Jim proved to be amazingly effective as an observation platform for a wide-eyed scientist entranced with the ability to walk around in a lively community of pink, gold, and bamboo coral, silver-gray eels, red crabs, and small black fish with glowing, luminous spots.

The experience provoked an enduring desire to see if new ways could be developed to go deeper, stay longer, and do more once there. In 1979, support was hard to come by for creating new technologies to explore the ocean with no particular objective other than to "see what's there." Eventually, in 1981, in collaboration with Graham Hawkes, a British engineer who had worked on Jim and developed several new systems for oil field diving, I founded—with lofty goals but scant funding—two new ventures, Deep Ocean Technology, and later, Deep Ocean Engineering. Initially, only the offshore oil industry appeared willing to support the technologies we proposed to improve working access in the sea.

Over the next several years, as president of the two small technology companies, I got to know many of the pioneers in the offshore oil and gas industry, and witnessed the transformation from something approaching a Wild West attitude about working on and in the sea to a far more sophisticated, technologically savvy industry, using the most advanced methods known for finding, extracting, and transporting oil and gas. Active platforms at sea currently involve technologies and protocols more akin to the space program than the rough-and-ready approach of the early years of offshore operations.

Worldwide, more than 4,000 rigs are now working in depths up to 1,830 meters (6,000 feet). The deepest offshore rig, Shell

Perversely spilled oil blackens Saudi Arabian marshes in 1991.

Oil's *Perdido Spar,* is operating in nearly 2,400 meters (8,000 feet) of water in the Gulf of Mexico. Standing as tall as the Eiffel Tower 350 kilometers (220 miles) offshore, the giant spar supports 150 workers and two long-range helicopters, and supplies meals worthy of a four-star hotel.

Largely because the action occurs underwater, out of public view, little attention has focused on what actually happens on the ocean floor where drilling takes place, or what creatures are displaced by the thousands of miles of pipeline laced across the bottom to transport oil and gas once they are extracted. Until the concern about the burning of fossil fuels—coal, oil, and gas—as the basis for increased carbon dioxide in the atmosphere, and thus global warming, ocean acidification, mercury contamination, and more, the greatest problem most people had with the development of offshore oil was the spilling of it.

THE ILLS OF SPILLS

A SeaWeb survey conducted by the Mellman Group in the mid-1990s found that "Americans believe the ocean's problems stem from many sources, but oil companies are seen as a prime culprit. In fact, 81% of Americans believe that oil spills are a very serious problem." Actually, spills *can* cause serious problems when they happen, and starting in the 1960s, they occurred often enough to arouse considerable public awareness—and ire.

Several mega-spills in the 1970s—*Torrey Canyon, Amoco Cadiz,* the uncontrolled gush of oil from Mexico's Ixtoc I well—resulted in a spate of national and international regulations aimed at improving conditions for producing and shipping oil. Meanwhile, thousands of less conspicuous but insidious mini-spills continued from oily bilgewater released from ships, oily water drizzling down street drains, and millions of small leaks and spills from boats around the world.

A pivotal shift in public attitudes about oil spills came in April 1989 when the tanker *Exxon Valdez* split when it went aground in Alaska's Prince William Sound, spilling 11 million gallons of Alaskan crude oil into the pristine waters. It dwarfed subsequent U.S. spills—the *World Prodigy* in Rhode Island; the *Presidente Rivera* in the Delaware River; the *Rachel B* / Coastal Towing barge collision in Galveston Bay, Texas, and *Mega Borg* in the Gulf of Mexico; *American Trader,* and later, M/V *Cosco Busan,* both in California; and a large but unnamed spill in Tampa Bay in 1993.

With as much scientific detachment as I could muster, in May 1989 I tried to evaluate the reality of the spilled Alaskan oil as I skidded over oil-slick beach boulders, dug into oil-saturated sand, held barely moving oil-coated crabs, looked out over oil-sheened waters, and listened to the wails of young otters, cleaned of oil but soaked in trauma as they huddled in holding cages at Valdez. My attempts to be open-minded collapsed as the horrendous toll continued to grow, and with it, a sense of despair about the nature of human nature, the indifference that caused the catastrophe, compounded by indifference that magnified what might have been a small spill into the largest in the history of the United States.

Outrage fostered by the *Exxon Valdez* spill centered on several key issues. In an article in *Science,* Eliot Marshall articulated one of them: "When the oil companies won permission from Congress in 1973 to lay a pipeline quickly from Alaska's North Slope to the port of Valdez, it was understood that ocean transport would be the riskiest part of the operation. . . . The potential disaster of a major spill in Prince William Sound was forecast—the bumbling response was not."

As it turned out, only about 4 percent of the oil spilled was recovered during the first critically important three weeks following the accident. Uncertainties about who was supposed to do what, and lack of mandated on-site equipment, delayed action when the oil was most concentrated and easiest to recover. Within hours, oil slipped over miles of calm water; after three weeks, it had slithered onto Alaskan beaches hundreds of miles away, smothering and poisoning the life out of hundreds of sea otters, thousands of seabirds, and billions of small creatures that quietly died as a consequence of human error exacerbated by human indifference.

The presiding federal judge for the *Amoco Cadiz* oil spill trial summed up the attitude of many with his comment: "The truth is we can never prevent oil spills. . . . They are the price of an oil-driven civilization just as deaths on the highway are the price of a mobile society."

But this time, public opinion was not quelled by the thought that spills are the inevitable cost of doing business. This was so clearly a preventable accident, the aftermath so fraught with needless confusion, the on-site preparation so lacking in compliance, and the consequences so devastating to wildlife and to Alaskans that strong regulatory measures, dreaded by the industry, simply had to follow. The 1990 Oil Pollution Act, passed quickly and easily by Congress after the tragedy, contains stiff safety provisions including a mandate for double-hull construction for oil tankers.

Regulatory action had no effect on the "mother of all oil spills," as some called the deliberate release into the Persian Gulf in 1991 of more than 45 times the amount of oil lost from the *Exxon Valdez.* Ecoterrorism is how some characterized the 500 million gallons of oil unleashed in an act of defiance by retreating Iraqis commanded by Saddam Hussein in Iraq's losing war with Kuwait. What followed was an eco-disaster for the people and wildlife of the entire region. While the majority of the oil flowed onto the beaches, reefs, and marshes of the upper Gulf, from Saudi Arabia to Kuwait, the lives and livelihoods of people and wildlife over a wide area were diminished or lost.

Checking out the biggest oil spill ever was not on my mind when I accepted the appointment as NOAA's chief scientist in 1990, but I soon became engulfed with U.S. efforts to track and evaluate the consequences of that devastating event. In my 1995 book *Sea Change,* I attempted to share the smell, look, and feel of 100 million barrels of oil swallowing Kuwait's once beautiful desert, now glistening in places with black lakes of oil that quickly entombed the life of the land—moths, dragonflies, birds, lizards, desert mice.

During repeated dives in the region from 1991 to 1993, I felt despair over the horrendous power of my species to destroy, knowing that the gulf and adjacent lands can never fully recover from the human conflicts that have ravaged the region in recent years. But I smiled when I saw new green shoots of marsh grass sprouting right through a sticky black shroud of oil, and laughed out loud in the harsh light of a burning oil well when I saw a line of ants working around a pale circle of sand, restoring order to their home, grain by grain. Biologist-philosopher-ant expert E. O. Wilson speaks, sometimes, of his concerns that we are allowing nature to slip through our fingers. Considering the resilience of nature, the greater concern, perhaps, is that we might be doing the slipping.

THE REAL PROBLEM WITH OIL

In 1982, I was among 90 participants from 38 countries invited by the United Nations Environment Programme (UNEP) to attend a two-day meeting in London to consider the state of the planet, and to identify the greatest threats to the world as we then knew it. A decade earlier, the UN Conference on the Human Environment had taken place in Stockholm, an event that resulted in the creation of UNEP. It seemed fitting that after ten years, an assessment might be in order.

Comments ranged widely, including a reminder from Mohammed Kassas, president of the International Union for Conservation of Nature, who spoke about food security, saying, "these resource ecosystems we call renewable . . . they are renewable if conserved, and they are destructible if they are not." Loss of soil to support agriculture was a major concern. The rapid and increasing rate of desertification was another, coupled with deforestation. Some were concerned about acid rain, the growing hole in the ozone layer, and the prevalence of diseases, especially malaria.

Norwegian explorer Thor Heyerdahl asked, "[Where] do we send all the pollutants . . . ? We are sweeping the floor and throwing it all under the carpet—and this carpet, the ocean, is the most important part of the planet." He added, "I am convinced that man today overestimates the size of the oceans and underestimates the importance of life on the planet." Russell Peterson, president of the Audubon Society, chimed in, "We are using the seas as dumps for our wastes, and mechanically destroying the nursery grounds for fish." Ocean explorer Jacques Cousteau added, "The fate of mankind has been tied up with the fate of water since the beginning of life." When it was my turn to speak, I said, "Weather and climate are shaped by the oceans. The oceans are home for the greatest diversity of life. If the oceans are changed, the character of the planet will be changed."

Meteorologist Robert White, NOAA's first administrator, identified climate change as a worry: "When we burn fossil fuels—carbon that has been laid down over the aeons—we are disrupting the natural balance. . . . Many nations plan big increases in the use of fossil fuel in the coming years. Yet the only practical thing which could slow down the carbon dioxide problem would be to burn less fossil fuel. We *must* push on with the development of renewable energy resources."

In summarizing the principal threats facing the world in 1982, however, the concern that rose above all others was the potential for devastation through the use of nuclear weapons. After that, if there was an "after," three closely connected topics were highlighted: overpopulation, poverty, and environmental degradation—that is, loss of species and loss of life-support functions.

A quarter of a century later, those issues remain, but Robert White's prescient concerns about climate change now underpin—and overshadow—all of them.

Fragmenting ice borders an Antarctic island.

CHANGING CLIMATE, CHANGING CHEMISTRY

We are already well into a new geological era, the Anthropocene, where human interference is the dominant factor in nearly every planetary ecosystem, to the detriment of perhaps all of them.

—Mark Lynas, *Six Degrees,* 2008

He had to be right. After all, he was the respected biology professor, and I was a student. The topic was whether or not humans can alter the planet's basic functions—temperature, rainfall, climate, atmospheric gases. I listened, but was not convinced.

His lecture continued:

> Earth is too large, humankind too insignificant. . . . There is no way that people building cities, turning forests into farmland, damming rivers, paving wetlands, burning oil, eliminating wildlife on the land or in the sea, could alter the nature of the world. . . . Things might not be as pretty, but the planet will hold a steady course, no matter what humans do—or don't do. . . . Big changes that do occur, such as the gradual shifts between one ice age and another, drought, floods—these things are caused by forces beyond anything humans can influence.

Perplexed, I said, "When I go to Philadelphia or New York, the air is different from the countryside 30 miles away. It looks brown, my eyes burn. It's always hotter in big cities, even in the winter. In the summer, you can't walk barefoot on the sidewalks and streets, but the grass is cool. Isn't it possible that all that cement, those roads and cars, the smokestacks, factories, and furnaces are making a difference, somehow? And aren't plants responsible for generating oxygen and taking up carbon dioxide? If we lose forests and marshes, doesn't *that* make a difference, too?"

"Show me the evidence," he said, not smiling.

Had I been less intimidated, I might have asked, "Can you show me the evidence that it is not?" But it was 1958, and like many others, I only had hunches, not proof, that what seemed like full-speed-ahead progress had some downsides.

THE EVIDENCE

It was the same year that geochemist Charles Keeling began taking precise measurements of carbon dioxide in the atmosphere from the rocky peak of Mauna Loa in Hawaii. Over time, the measurements yielded evidence that the rate of CO_2 in the atmosphere has climbed from the pre-1800, preindustrial-age level of 275 to 280 parts per million (ppm), to 315 ppm when Keeling began his research, and then to 385 ppm half a century later.

It doesn't sound like much, but scientists checked it against the amount of CO_2 found in ancient air trapped in cores of Antarctic ice. They discovered that 275 to 280 ppm held steady for millennia: The present level is greater than it has been in 800,000 years. All arrows are pointing to a continuing, sharp, worrisome increase.

Now we know not only that the level of CO_2 is increasing, but also that there are clear causes and measurable consequences.

Before the middle of the 20th century, the technologies did not exist that would enable humankind to understand that we do have the power to fundamentally alter the way the world works. Now there is *evidence,* with increasing CO_2 just one of the many troubling ways we unwittingly have been chipping away at the underpinnings of planetary stability.

Those "natural forces beyond human control" that my old professor relied on to explain shifts between ice ages and warm periods: the wobble and shift in Earth's orbit and the tilt of its axis? Alone, they do not explain the recent rapid upswing in temperature. They are overwhelmed by the influence of greenhouse gases.

Our species has two ways of influencing the increase of carbon dioxide in the atmosphere, and in turn, the upswing in temperature. First, we are releasing into the atmosphere huge amounts of carbon dioxide through the combustion of coal, oil, and gas. In just decades, we are consuming as fuel ancient forests and the essence of ancient ocean microbes compressed over hundreds of million of years. The World Meteorological Organization's volume *Climate: Into the 21st Century* succinctly sums up the problem and the solution: "Emissions due to burning fossil fuels will be the dominant factor in the buildup of carbon dioxide concentrations in the 21st century. Reducing carbon dioxide emissions is the most important factor for lowering the global impact of human activities on the climate system."

Currently, up to 20 percent of human greenhouse gas emissions are being caused by deforestation in tropical Brazil and Indonesia, making those countries two of the highest carbon emitters in the world. It is estimated that halting forest destruction would save the same amount of carbon over the next century as stopping all fossil-fuel emissions for ten years. But at the present time, forests are valued primarily for their timber, not their vital role in biodiversity or as carbon sinks.

Additional carbon dioxide enters the air through the extraction of hundreds of millions of tons of fish, oysters, clams, and other carbon-based forms of life from the sea. Much of their carbon escapes back to the atmosphere when the energy stored within the animals is burned by the metabolism of the consumers of fish, fish meal, fish oil, clam chowder, shrimp cocktails, sushi, or sashimi.

The second way we are increasing the amount of carbon dioxide in the atmosphere is by decreasing the capacity of living systems, the biosphere, to use it. Fewer plants means less photosynthesis, and thus less carbon dioxide uptake. When we pave over the land, we are taking a great green wedge out of Earth's life-support system.

The destruction of forests to plant houses, farms, and shopping malls; the conversion of wetlands to condos and parking lots, sea grass meadows and mangrove forests to shrimp and fish farms, and coral reefs to road-building materials—all reduce the ability of Earth's living systems to consume carbon dioxide and generate oxygen. In the open sea where most of Earth's photosynthesis takes place, many of the photosynthetic microbes have lives measured in days, weeks, or months. Carbon in the sea is stored not in centuries-old trees, peat, and soil, but rather, in the steady rain of small creatures falling to the seafloor and by entering the food chain, moving into long-lived corals, sponges, mollusks, and others, and then into the decades- or centuries-old top predators that we continue to focus on for extraction. Killing turtles, whales, sharks, tunas, orange roughy, hoki, monkfish, sea bass, rockfish, groupers, cod, Arctic "cod," Antarctic "cod," and other long-lived species puts carbon dioxide back into the air and disrupts the capacity of the ocean to hold carbon in its system. Industrial fishing, in effect, has been clear-cutting ancient ecosystems, disrupting and dismembering the underpinnings of the dynamic but amazingly stable carbon cycle constructed over hundreds of millions of years.

THE WARMING PLANET

The recent rise in Earth's surface temperature is closely coupled with the correspondingly recent increase in carbon dioxide. Data gathered from thousands of locations across the planet, land and sea, show an increase of 0.74°C (1.3°F) in the past century, with the greatest increase occurring in the past 30 years. A report of findings of hundreds of the world's leading Earth scientists, the 2007 Intergovernmental Panel on Climate Change (IPCC), not only confirmed the warming trend but also unequivocally linked the warming to human activities. A one-degree change from dawn to dusk is barely noticeable, but on an overall planetary scale, a rise of that magnitude translates to shifts in what kinds of creatures can live in what areas; it can modify climate and weather patterns. Warmer air also increases the amount of water that can be held in the atmosphere, enhancing the effect of water vapor as a "greenhouse gas" that helps to seal in heat.

Carbon dioxide is vital for photosynthesis, and the normal level—0.03 percent of atmospheric gases—is just right to maintain productivity on the land and in the sea. Sunlight plus carbon dioxide and water in the presence of chlorophyll yields oxygen and the simple sugar that in turn provides the basis for carbohydrates, fats, and proteins that feed us and much of the rest of life on the planet. Excess carbon dioxide escapes into the atmosphere or is absorbed into the sea. Without life to absorb and transform carbon dioxide, Earth's atmosphere would most likely resemble that of Venus and Mars—more than 95 percent carbon dioxide, a trace of oxygen, about 2 percent nitrogen, and a smidgen of argon. The surface temperature would be far too hot to suit us —somewhere above 250°C (482°F).

It has taken about four billion years for living systems, mostly in the sea, to transform the lifeless ingredients of early Earth into the Eden that makes our lives possible, and less than a century for

Times are changing for polar bears and other creatures of the Arctic.

us to destabilize those ancient rhythms. Present climate change policies focus on the atmosphere, largely neglecting the ocean, despite ample evidence that the ocean drives and regulates planetary climate, weather, temperature, and chemistry. Grant Bigg, in *The Oceans and Climate,* notes: "[Oceans] store immense amounts of energy for months, decades or even centuries, depending on the region, depth and the nature of the interaction between the atmosphere and ocean. This storage capacity acts as a giant flywheel to the climate system, moderating change but prolonging it once change commences." Richard Spinrad, Assistant Administrator of NOAA's Office of Oceanic and Atmospheric Research (OAR) says, "The seas devour more than 22 million tons of CO_2 daily . . . they store 50 times more than the atmosphere."

Over time, global temperatures rise and fall, ice ages come and go, sea level increases, then diminishes. About 55 million years ago, when Earth's temperature was about five degrees Celsius (nine degrees Fahrenheit) higher than at present, forests, not ice, ranged far north and far south. Ice ages have come and gone many times since then, but conditions have remained remarkably congenial for humankind since the height of the most recent epoch, the Holocene, starting about 11,000 years ago—the time during which most of civilization has developed and prospered. Over thousands of years, sea level has ever so gradually been increasing, submerging regions such as Florida's west coast. Another Florida extends seaward more than a hundred miles, once home for mastodons and giant sloths, now populated by dolphins and groupers. In due course, today's shoreline hotels, restaurants, and condominiums will be homes for sea creatures.

MELTING ICE AND RISING SEAS

As long as carbon dioxide levels keep increasing, so will the temperature, and with it, the inexorable rise of the sea. The recent climb in temperature, slight as it might seem, has already triggered a significant loss of polar ice and a small but accelerating increase in sea level. As water warms, it also expands, a phenomenon that, coupled with melting ice, leads to about a 3-millimeter (0.12-inch) rise per year. That's not much to worry about if the rate holds steady, but projections of what is possible by midcentury are not comforting to those with waterfront property or island homes.

Aboard the Russian nuclear-powered, 213-meter-long (700-foot) icebreaker *Sovietsky Soyuz,* I went to the North Pole in 1998, when solid ice covered most of the region north of the Arctic Circle. Eighty percent of the heat in sunlight striking the brilliant white surface of the frozen Arctic Ocean is reflected away, but dark, exposed ocean water absorbs 95 percent. More water means faster warming that melts more ice, yielding more open water.

Standing on the icebreaker's deck, I watched the bow crack apart and flip great frozen slabs, slick on the underside with golden brown diatoms, home for pink crustaceans and an occasional silver fish. Smashing through ice as much as 4 meters (13 feet) thick, the ship left a wide swath of shattered fragments floating in a slushy path of dark water. I wondered if the pieces would soon refreeze, or if the crushed jumble in the ship's wake would remain an open river. Solid ice melts more slowly than crushed ice, and every tear in the Arctic Ocean's frozen ceiling means another small nudge toward dissolution. I was not aware at the time that Arctic sea ice has been in constant retreat since 1980, and may soon disappear altogether in summer months. Currently, the Arctic is warming twice as fast as the rest of the planet, and the pace is picking up.

At the other end of the planet, huge ice shelves surrounding the Antarctic continent have thinned and broken away in the past several decades. Satellite images of the Larsen B shelf in western Antarctica from January 2002 to March 2002 show the swift change from what appeared to be a solid mass marked with puddles and streaks of surface melt to the complete disintegration of an area the size of Rhode Island. According to a 2008 National Snow and Ice Data Center study of the 2002 collapse, rifts in the ice sheet had been growing for about two decades, the ice thinning and under pressure as glacier flow began to increase. Warming of deep Southern Ocean currents flowing along the Antarctic coastline appears to have eroded the ice shelf from below, making it more fragile. The unusually warm summer of 2002 nudged the already weakened mass of ice to the point of rapid dissolution.

The loss, while alarming, made an enormous region of the Antarctic seabed accessible to exploration for the first time, an opportunity seized by 52 scientists from 14 countries in a Census of Antarctic Marine Life expedition in 2006–07 aboard the research vessel *Polarstern.* Some imagined that there might not be much life under a frozen pavement of ice that had persisted for thousands of years, but the diversity discovered ranged from fields of glass sponges and giant sea stars to what appear to be new species of gelatinous mid-water animals. Polar biologist Gauthier Chapelle observed, "This is virgin geography. If we don't find out what this area is like now following the collapse of the ice shelf, and what species are there, we won't have any basis to know in 20 years' time what has changed, and how global warming has altered the marine ecosystem."

In 1990, I worked with a team of scientists using a remotely operated vehicle to look at the seafloor under the ice in the Ross Sea near McMurdo Sound. We worked close to the base most of the time, but occasionally left the little submersible behind and traveled by helicopter to the ice edge. From the air, the ice shelf looks as solid as marble, a glistening alabaster plain bordered by deeply indigo water. But standing close to the edge of the Southern Ocean, I could feel the insistent pulse of the sea below, gently lifting the ice, then falling away, like gentle breathing.

Several feet of ice were under my heavily insulated boots, and under that more than 305 meters (1,000 feet) of the coldest ocean water there is—as low as minus 1.8°C (29°F). To humans, this is an extreme environment, but to the creatures that live there, New York City would be an extreme environment, as would any place other than the comforts provided by dark, near-freezing water. Near the surface, golden brown diatoms abound during summer months, as well as enormously rich and diverse groups of micro-animals, from shrimplike krill and swimming pteropod snails to much smaller, less conspicuous grazers and legions of mini-predators.

Deeper down, in eternal darkness, a unique family of fish thrives. The Nototheniidae—commonly known as ice fish or antifreeze fish—include the newly popular "Chilean sea bass" (properly called Patagonian toothfish) and Antarctic cod (properly called Antarctic toothfish) among the 60 or so species that are of great interest to those studying human disorders ranging from osteoporosis to blood diseases. Unique among vertebrates, ice fish lack hemoglobin and red blood cells, but their body fluids contain "antifreeze" proteins that inhibit the formation of ice crystals. They also lack a swim bladder and have exceptionally lightweight skeletal systems. In the 1970s, industrial fishing by the U.S.S.R. seriously depleted Antarctic ice fish, and fleets from several nations continue illegal, unregulated large-scale fishing, putting the future of these amazing creatures at great risk—especially when coupled with the changes to their deep-ocean realm caused by global warming.

As I stood at the edge of the Antarctic Circumpolar Current, the largest moving mass of water on the planet, it was hard to imagine how one terrestrial species (that would be us) could in so short a time have so profound an impact on so many that have endured so long. Several minke whales cruised by, first visible as exhaled puffs fogging the air, jetting from heated bodies as warm as mine, fueled with sunshine concentrated in the legions of phytoplankton engulfed by the krill they in turn had consumed.

A small band of Adélie penguins popped out of the dark water like sleek black-and-white corks and sprinted full speed ahead, rocking slightly from side to side as they padded straight for us. A few feet away, they stopped and simply looked. I want to say they *gawked,* because that's what it seemed like: curious youngsters staring at strange creatures that just had to be investigated. Curiosity satisfied, they wandered off until they were small black specks, plunging back into the sea. Several of their black-tie–attired cousins, emperor penguins, slipped onto the ice with no apparent effort and approached with a more circumspect, dignified, but equally curious demeanor. After giving us intense scrutiny for half an hour or so, they, too, slid back into the sea.

At the time, I had no idea that Antarctic ice could melt as fast as it has in recent years. I knew that 90 percent of Earth's fresh water is there, in massive sheets covering 98 percent of a continent half again as large as the United States. I had heard that if all of the ice in Antarctica were to melt, sea level would rise more than 60 meters (200 feet), as it has done repeatedly in the past—but at a geologically stately pace. Now I understand that human impacts on global climate are creating change literally of geological magnitude in a few decades. The penguins, whales, or krill may sense that their world is changing, but if they do, they can't know why or do anything to modify their fate. We can do both but seem to be moving at a geological pace ourselves to take actions that can make a difference.

METHANE AND LIFE

Numerous models have been crafted to anticipate the consequences of increasing CO_2, with the conclusion that a 500 ppm concentration would most likely warm the planet by 3° to 4°C (5.4° to 7.2°F). The permafrost underlying the Arctic tundra, already softening, would generally melt, giving up its stored water and methane, greatly enhancing the greenhouse effect and accelerating the warming trend. Although methane dissipates faster than carbon dioxide once it is released into the atmosphere, during the decades that it is there it is about 20 times as potent as carbon dioxide. At present rates, carbon dioxide is expected to reach 500 ppm by 2050, and combined with the effects of methane and nitrous oxide, it will cause a relentless push toward higher temperatures, which in turn will release more methane from tundra regions, increasing warming in a relentless feedback cycle.

Less obvious but potentially a greater cause for concern than gases within the Arctic tundra is the enormous accumulation of methane deep within the sea. Where oil and gas deposits occur, methane gas emerges. Beginning hundreds, sometimes thousands of feet underwater, the gas forms tiny spheres that increase in size as pressure decreases, rising within the water column until they finally burp into the atmosphere.

Much greater quantities of methane—possibly more than all other petrochemicals combined—are locked away in the deep sea as gas hydrates or clathrates. These masses of methane are caged within latticelike ice structures and held in place by the pressure and cold temperature characteristic of the deep sea below about 600 meters (2,000 feet). Fishermen sometimes catch chunks of methane hydrate in their nets when trawling in deep water, and they get a firsthand view of the release of methane into the atmosphere when the ice disintegrates into a puddle of water amid much sizzling and popping: methane escaping into the sky. With continued warming, the sea may give up in a geological blink the accumulation of many millions of years of carbon sequestration. Destabilizing the massive accumulation of methane hydrates could trigger undersea landslides that in turn could set off tsunamis on a grand scale.

During the 2001 Sustainable Seas Expeditions, while piloting the little *Deep Worker* submersible 161 kilometers (100 miles) offshore from the mouth of the Mississippi River and 550 meters (1,800 feet) down, I witnessed streams of methane glistening silver as they bubbled from the silty seafloor. At the time, I was intent on locating a deep reef of the coral *Lophelia* that had been discovered by someone dragging a net across the bottom in deep water years before. I did not locate the reef, but fortuitously landed right in front of a great interwoven mound, what appeared at first to be a pile of dead sticks or the abandoned nest of a giant underwater bird.

I knew right away that, happily, I had encountered a thriving community of tube worms. Sometimes called gutless bearded worms, since they have no mouth or digestive system, and some have tentacles resembling whiskers, tube worms are creatures whose large, red-plumed cousins were first discovered near hydrothermal vents three kilometers (two miles) deep near the Galápagos Islands in 1977. Named *Riftia,* the lanky, 2-meter-tall (6.5-foot) animals in Galápagos waters literally revolutionized several fields of science when further research led to the discovery of microbes—more than a hundred billion living in a spoonful of the tube worms' tissues—that use chemicals from the surrounding hydrogen sulfide–rich seawater to generate sustenance through chemosynthesis. One hundred *billion* individual morsels of life in a spoonful gives perspective to the concept that on this planet, without a doubt, microbes rule.

In the absence of sunlight, bacteria, as well as the newly discovered kingdom of microbes known as the Archaea, use hydrogen sulfide (a poisonous gas that smells like rotten eggs), carbon dioxide, and oxygen to form nutritious organic carbon compounds. Heat was thought to be essential for the process because the initial concentrations of life were found around hot-water volcanic vents in the seafloor, but in the 1980s, similar forms were found around *cold* seeps of methane gas in the Gulf of Mexico.

Consider the vast area of the world that is under the ocean, all of it dark some of the time, and most of it dark all of the time. Various forms of chemosynthesis are taking place here and there within the enormous liquid space between the photic zone and the deep-sea floor, as well as *across* the seafloor, within the sediments *on* the seafloor, and in the water-filled cracks *beneath* the seafloor. It is no wonder that some speculate that the amount of carbon taken up through chemosynthesis in the darkness of the ocean may rival photosynthesis in terms of fixing, storing, and passing along carbon through intricate food webs.

This thought is more than a hunch and less than a tested idea currently, but given our enormous ignorance about what lives in the ocean, and given the real possibility that such thoughts may soon emerge as cutting-edge insight, the concept bears consideration. At the very least, chemosynthesis needs to be considered as a factor in the climate-change equation—even if it is an *x* for "unknown" on the planetary carbon balance sheet. Generally speaking, forests tend to be accounted for, and sometimes the role of photosynthesis in the well-illuminated surface waters of the sea, but recognition of the importance of the deep-sea relative to the carbon cycle and climate change rarely gets beyond the laboratories of the scientists who work there.

Enormous interest has been aroused by the possibility that Europa, one of the moons of Jupiter, might harbor life in the dark seas believed to exist below several miles of ice—life that might use the magic of chemosynthesis to power food webs. Meanwhile, chemosynthesis occurring on this very up-close-and-personal planet is being given relatively short shrift. It is vital that we look for the possibility of life elsewhere in the solar system and the universe beyond, but it also seems logical to explore the tantalizing mysteries awaiting attention in this part of the universe.

Consider the haunting possibility that bioluminescence or geothermal light might provide just enough illumination to initiate photosynthesis far below the depths where sunlight penetrates. In a recent discovery made more than 2,000 meters (6,600 feet) deep at a hydrothermal vent on the East Pacific Rise, in water registering 350°C (662°F), faint geothermal light appeared to trigger a previously unknown kind of photosynthesis involving green sulfur bacteria. I have personally puzzled over the occurrence of bright green tissues deep within the limestone structure

An uncertain future awaits penguins and other Antarctic life.

of coralline rock 200 meters (660 feet) under the sea, and gazed in wonder at green algae living on the underside of rocks in Hawaii in nearly 30 meters (100 feet) of clear ocean water where the world is distinctly, comprehensively blue.

Such thoughts were not on my mind while staring through the clear dome of the *Deep Worker* submersible far beneath the ocean's sunlit surface. I was focused on the delight of seeing the Gulf's tube worms and columns of methane streaming upward through the tangle of creatures so recently discovered, so exciting in terms of the impact they were having on the scientific community, if not the world at large.

I tried to imagine the reaction if, when I returned, I accosted someone in my neighborhood supermarket and said, "I have just seen the most wonderful thing! One hundred miles offshore from the mouth of the Mississippi River, 1,800 feet under the sea, there are spider crabs and bright red shrimp, small red 'gaper' fish with great, wide mouths, scarlet starfish and shrimp with topaz eyes glittering among the branches of these ten-foot-tall creatures known as gutless bearded worms that have trillions of bacteria living in their tissues, grabbing chemicals out of the water—including the carbon dioxide that is a real problem associated with global warming and . . ."

Well, there's the rub. How is it possible to convey in words that mean anything to anybody the serious state we find ourselves in? A few scientists know. The information is there. What will it take to convey to all who need to know that we are undermining the capacity of the planet to support the life we know and love. Our future is in jeopardy, while we fiddle with this and that, not willing to embrace the *evidence*. Meanwhile the microbes wait, and will continue to do just fine, no matter what.

GOODBYE, ARCTIC ICE?

One scientist who knows how to convey this information is the tall, lean, mountain-climbing artist-scientist James Balog, who not only has figured out how to photograph redwood trees in their top-to-bottom entirety, but also puts himself and teams of equally "out there" camera-wielding colleagues on mountain peaks and icy crevasses to document the changing face of the planet's ice. The time-lapse photography from Balog's project, the Extreme Ice Survey, should be shown to everyone as a required reality check on the astonishingly swift—and widespread—disappearance of ice everywhere from mountaintops to glaciers to Greenland's icy fringe.

Ask Balog about the consequences of the fast-track warming trend we are now experiencing and he will most likely look you in the eye and say something close to what he wrote in his slim 2009 volume *Extreme Ice Now:*

> To varying degrees in different parts of the planet, [warming] dries up water supplies, disrupts agriculture, increases prices of food and water, makes extremes of weather more extreme, raises sea level, sets off more wildfires that burn more intensely, spreads mosquito- and rodent-borne disease, causes the extinction of some animal and plant species, and helps undesirable animals and plants to proliferate. . . . According to the best current estimates, sea level will rise by at least two feet, probably three feet, and possibly four to five feet by 2100. . . . Whatever happens, the human race will have an interesting future.

According to present projections, we have less than a decade to cut back global emissions to a level—under 400 ppm—that could stabilize the warming trend: a daunting but not impossible goal. However, owing to the momentum already under way, even if all anthropogenic greenhouse emissions were to stop, the temperature would continue to rise for a while, by as much as half a degree.

Even if carbon dioxide and methane releases were held at their present levels, the processes now in motion are expected to yield an

Every drop counts: Melting ice results in rising seas.

ice-free summer at the North Pole by 2040. Given present trends, the disappearance of the entire Arctic ice cap soon may be inevitable, but NASA climatologist James Hansen offers hope. He believes that the complete loss of the ice may be prevented with a significant reduction in greenhouse gas emissions, coupled with a sharp reduction in atmospheric pollutants, especially the rain of small but potent particles of soot that darken the ice and accelerate melting.

As the rookie chief scientist at NOAA, the agency that oversees atmospheric as well as ocean issues, I had a glimpse of what soot-darkened snow and ice look like during a visit to the northernmost U.S. weather station in Barrow, Alaska, in February 1991. Drifts buried the main building except for a small portion of the roof and various stacks and pipes, protruding from what appeared to be a diamond-brilliant blanket of fresh snow. Tunneling to the door, I was welcomed into a warm nest of desks, tables, and instruments manned by two hardy scientists charged with monitoring the things you'd expect—temperature, humidity, snowfall, wind, windchill, dew point, pressure—and soot.

I was intrigued with the concept of air currents behaving much like ocean currents, picking up and conveying aerial debris over the top of the world, dusting distant drifts with hitchhiking particles belched from factory smokestacks in northern Europe and Russia. Each dark speck was like a fingerprint, traceable to its point of origin. Trillions of the minuscule bits of soot floating from the sky had succeeded in reducing the reflectivity of the Arctic snow in recent years, each mote playing a tiny part in warming the planet. The massive use of coal as a source of energy since the end of the 1800s not only released great quantities of CO_2 but also powdered the atmosphere with an infinite number of small black dabs of soot. Coal continues to be a dominant energy source in the U.S., China, and many other countries, and its use is increasing.

The melting of the snow and ice that blanket the Arctic Ocean does not increase sea level, because that ice is already part of the sea's volume, but water flowing from the melting ice that caps Greenland and other landmasses certainly does. Satellite measurements as well as monitoring sites on the ground confirm the retreat of glaciers at a speed much faster than was predicted even a decade ago. Melting ice sinks through cracks and lubricates the bottom of glaciers, sometimes hundreds of feet below, speeding their slide to the sea. Rivers of meltwater are now plunging through icy glacial sinkholes, called moulins, onto bedrock beneath the ice sheet.

The changes now taking place are bad news for the cultural traditions of the people of the Arctic as well as the polar bears, bowhead whales, narwhals, seals, and other creatures whose lives are utterly dependent on the integrity of an ice-capped Arctic. Within decades, not millennia, shifts are occurring that might gradually be accommodated given centuries to adapt, but are disastrous when happening as abruptly as they are.

Abrupt climate change, once thought to be in the realm of science fiction, is already here, and could soon become abruptly obvious. Most climate change models show a weakening of the ocean's circulation, largely driven by temperature and salinity, and a reduction of ocean heat transport into the high latitudes of the Northern Hemisphere. In short, that means the possible diversion southward of the Gulf Stream from its present sweep across the Atlantic where it now mellows what would otherwise be a much colder England and other northern European countries.

AN ACID OCEAN

Only about half of the annual release of six billion metric tons of CO_2 from burning fossil fuels increases CO_2 in the atmosphere. The remainder goes into the ocean and is giving rise to an unexpected suite of additional concerns. Of greatest worry is the steady rise in ocean acidification caused by the conversion of the excess carbon dioxide into carbonic acid: As the amount of carbon dioxide in the ocean increases, so does the water's acidity.

Some think first of the consequences to coral reefs, where increasing acidity can interfere with the ability of corals to build and maintain their stony skeletons; past a certain level, they dissolve. The structure of coral reefs also depends on red and green species of coralline algae that may make up as much as 90 percent of the mass of a coral reef. They, too, dissolve when the surrounding water becomes too acidic. Everything with a calcium carbonate shell is vulnerable—oysters, clams, snails, pteropods (planktonic swimming mollusks), many sponges, sea stars, sea cucumbers, sea urchins—the list is long. A change in acidification can cause trouble for everything from developing fish to jellyfish. Alter the chemistry of the ocean, and the entire system shifts. Some natural changes we can predict, but it is impossible to anticipate how fast, or how much will occur as a consequence of tipping the ocean's chemistry onto a different course.

We have real concerns about clear-cutting and burning forests across the globe because of the resulting release of carbon dioxide and interference with the natural systems that usually absorb carbon dioxide from the atmosphere. But the microorganisms in the sea extract more carbon dioxide than photosynthetic organisms on the land, and if they are shut down through acidification or other factors, the loss of the natural capacity of the planet to take up carbon dioxide is proportionally diminished.

Most worrisome is the effect rising acidification is having on the very small photosynthetic organisms that generate much of the oxygen in the atmosphere. Trees, grasses, and other land plants are critically important in terms of maintaining the atmospheric gases in just the right proportion suitable for present life on Earth, including us, but photosynthetic organisms in the sea do most of the heavy lifting when it comes to generating oxygen and otherwise holding planetary chemistry on a steady course. As acidification increases, acid-tolerant organisms will prosper, and some now present in small numbers are likely to increase. Those that require the alkaline environment that has characterized the chemistry of the ocean for millions of years will fade.

Among the organisms most vulnerable to the present rise in acidification are coccolithophorids, minute photosynthesizing protists with intricate, lacey, limestone shells, and foraminifera, single-celled animals with exquisitely beautiful shells that make up enormous deposits of calcium carbonate on the seabed. They are also an important component of ancient limestone rock, such as the white cliffs of

Dover, the underpinnings of Florida, and other conspicuous limestone deposits around the world that once were under the sea.

Mixed with the many profound, deep causes for concern are profound, deep reasons for optimism. Robert Socolow a professor of engineering and Stephen Pacala, an ecology professor, both from Princton University, say, "Humanity can solve the carbon and climate problem in the first half of this century simply by scaling up what we already know how to do." To meet the goal of maintaining something close to present levels of atmospheric carbon as populations and industry grow requires reducing carbon emissions by one billion tons by 2055.

To achieve this, some, including the visionary biologist and big-picture thinker James Lovelock, favor a greatly scaled-up and carbon-free use of nuclear energy. Others dream of engineering a way to improve upon what plants do naturally—creating artificial photosynthesis. Some aren't waiting, but are already clearing forests to plant energy-rich crops such as corn to produce biofuels, while still others are working with fast-growing microbes in closed systems to achieve similar clean-burning fuel. Clearing forests to plant corn adds carbon to the air rather than removing it, but there may be hope for contained cultures of certain microbes in terms of generating biofuels efficiently.

Wind turbines, much more efficient editions of the tried-and-true windmills of times past, are becoming adopted widely, while vastly improved solar arrays are revolutionizing clean energy. Eventually, solar power will be likely to dominate the spectrum of energy sources required to accommodate the present and future needs, if not the desires, of the billions of people who now embrace the world. Storing and distributing power once generated remain as major problems in need of swift solutions.

All things considered, it would have helped enormously had we known and taken seriously 50 years ago what now is apparent concerning climate change. We cannot go back, of course, but we should do our best to keep our options open, our goals for success up—and our emissions down. Jim Balog provides useful perspective: "Unless we take action, today's opportunities will be tomorrow's crisis."

Balog shares with thousands of scientists around the world a profound sense of urgency laced with frustration over the slow response of the public and officials to take actions that can stabilize and reduce carbon dioxide levels.

The crisis of today is one of complacency. Although 450 ppm and a two-degree rise in temperature appear acceptable to some, the last time Earth warmed to that level, sea level rose by tens of meters and climate was dramatically different from what we have today.

NASA climatologist Jim Hansen wrote in an abstract for an article submitted to *Science* magazine in 2008, "If humanity wishes to preserve a planet similar to that on which civilization developed and to which life on Earth is adapted, paleoclimate evidence and ongoing climate change suggest that CO_2 will need to be reduced from its current 385 ppm, to at most 350 ppm." To enlist public awareness and support for action, a promising effort called "350.org" is being spearheaded by author-environmentalist Bill McKibben. While serving on a climate change panel at a scientific conference in New York with McKibben and Hansen in June 2009, I was reassured that the 350 goal is not only achievable but may be mandatory to avoid truly cataclysmic consequences.

As a parent and grandparent, I share the rationale underlying Hansen's personal sense of urgency, quoted in a 2009 *New Yorker* profile: "I decided that I didn't want my grandchildren to say, 'Opa understood what was happening, but he didn't make it clear.'"

III. NOW IS THE TIME

OPPORTUNITIES FOR ACTION

Millions of divers now explore the world from the inside out.
PREVIOUS PAGES: *The southern ocean is home to these Adélie penguins.*

EXPLORING THE OCEAN

8

Most people walking around in a mall or on a college campus are carrying on them better technology than the entire U.S. government had when it put a man on the moon. Each one of us is a walking technological superpower. . . . Given the capacities available to us, our wildest dreams and biggest hopes are probably too small.

—Van Jones, *The Green Collar Economy,* 2008

Reasons for exploration abound—acquiring territory, seeking wealth, looking for new ways to get from here to there—but above all, it comes down to curiosity. Throughout history people have been lured to new lands, climbed mountains, sailed over the surface of the ocean, reached for the moon and actually gotten there in large measure because of the very human need to know, to understand. T. S. Eliot is often quoted for his enigmatic way of expressing this incurable itch in *Little Gidding,* the last of his *Four Quartets:*

We shall not cease from exploration
And the end of all our exploring
Will be to arrive where we started
And know the place for the first time.

Philosopher-scientist E. O. Wilson, at his 80th birthday celebration in New York in 2009, told the assembled crowd of well-wishers that humans have always been driven by great questions: Who are we? Where did we come from? What is our future? Why are we here? How can we find out unless we explore, and continue to do what most of us tend to do as children: ask how, when, why, where, what, laced with liberal doses of open-eyed, open-minded, open-hearted unabashed sense wonder—then go look for answers?

Early ocean explorers had these attributes but were limited by the technologies available to probe the depths. Even today, scientists rely heavily on nets, dredges, and remotely deployed devices to sample and document the nature of the deep sea. Healthy humans can trudge on foot to the deepest forests, driest deserts, and iciest glaciers; ascend the highest places above sea level. Going more than a hundred feet or so below the ocean's surface is another matter.

Ignorance about the ocean and lack of understanding about the relevance of the fundamental ways people everywhere are connected to the sea underlie the indifference and complacency that have led to abuses imposed by people in recent years, with dire consequences to our future. In the 20th century, tremendous investments were made in aviation and space technologies that gave humankind ready access to the skies above and advanced knowledge of the heavens beyond. Those investments have paid off mightily. Meanwhile, we have neglected the ocean—and it has cost us dearly.

There are problems to be solved if we are to be able to explore the depths of the sea, of course, but it was also tricky getting to the moon and sending probes to Mars and Jupiter.

GETTING THERE

Free diving on a single breath of air, a few hardy humans have descended to more than 150 meters (500 feet) for brief excursions into the twilight zone—that softly illuminated, dusklike region in the sea topping the ever dark realm that is home for most living things. To be able to spend meaningful time or even to send effective cameras and sensors below a few hundred meters requires the use of creative technologies.

Commercial divers began using "heavy gear" in the 1830s—cumbersome suits, lead weights, lead shoes, and metal helmets connected by a hose to surface-supplied compressed air—to build tunnels; salvage wrecks; conduct military operations; look for treasure; and collect sponges, pearls, and precious corals. But one of the first scientists to use such gear for exploration was naturalist William Beebe, who adapted a copper helmet to explore coral reefs in Bermuda in the late 1920s. He suggests, in his account *Beneath Tropic Seas:* "Don't die without having borrowed, stolen, purchased or made a helmet of sorts, to glimpse for yourself this new world. Books, aquaria and glass-bottomed boats are, to such an experience, only what a time-table is to an actual tour. . . ."

I took Beebe's words to heart, and while still a teenager, tagged along with my brother and a friend whose commercial-diver father probably was unaware that we had "borrowed" his helmet and compressor for dives in Florida's Weeki Wachee River, the first dives any of us had had using compressed air.

The development of the system scuba divers use today got under way in 1942 when the French Navy captain Jacques Cousteau met with engineer Emile Gagnan to discuss how to devise a way to breathe underwater from a compressed-air tank. As a result of their deliberations, a demand valve Gagnan had developed to feed gas automatically into the motors of automobiles morphed into the first automatic regulator for diving.

Cousteau wrote of his first experiences as a "fish man" with his new "Aqualung," in a 1952 article in *National Geographic:* "The

Aqualung frees a man to glide, unhurried and unharmed, fathoms deep beneath the sea. It permits him to skim face down through the water, roll over or loll on his side, propelled along by flippered feet. . . . In shallow water or in deep, he feels its weight upon him no more than do the fish that flicker shyly past him."

I was among the many who could not resist the appeal of Cousteau's descriptions of "flying" underwater, and using scuba in the 1950s, I explored the coastline of the Gulf of Mexico from the Mississippi Delta to the tip of the Florida Keys, often diving alone before the obvious benefits of "buddy diving" became routine. In classes at Duke University in the 1960s, I tried to explain to my fellow students that using scuba, I had only 20 minutes to explore a reef in 30 meters (100 feet) of water before I had to return to the surface. Stopwatch science surveys, I called them. I wondered how much would be known of forests or mountains or archaeological sites anywhere on the land if you could only go 30 meters from your jeep with less than half an hour to look around!

At the time, enthusiasm for exploring space—even going to the moon—was paralleled by growing interest in accessing the ocean. In 1960, the same year that Yuri Gagarin became the first human to view Earth while in orbit above it, U.S. Navy lieutenant Don Walsh and Swiss oceanic engineer Jacques Piccard successfully viewed the world from below, from the greatest depth in the sea. While they demonstrated the ability to travel as deep as anyone could possibly go in the ocean, others were solving the problem of how to stay longer—to actually live underwater.

LIVING UNDERWATER

Jules Verne dreamed of staying submerged in *Twenty Thousand Leagues Under the Sea,* in 1870. In a conversation between the fictional Captain Nemo and Professor Aronnax, Verne wrote, "You know as well as I do, Professor, that man can live under water, providing that he carries with him a sufficient supply of breathable air. . . ."

A U.S. Navy captain and physician, George Bond, affectionately known as Papa Topside, was convinced that Verne's visions could be realized. In the 1950s, he figured out that "once a diver's body was saturated with compressed gas . . . the decompression time would be the same whether the diver stayed underwater for a matter of hours, days, weeks or even months. The amount of time necessary for decompression depends on the depth of the dive and on the gases breathed."

Bond experimented with his concept for five years in U.S. Navy laboratories before sharing his ideas with Jacques Cousteau and with aviation pioneer and engineer Edwin A. Link. By 1962, Cousteau and Link had independently developed successful saturation diving projects that provided the basis for dozens of subsequent systems, from the U.S. Navy's Sealab and other military systems to numerous commercial diving operations in support of the offshore oil and gas industry.

In 1969, while astronauts walked on the moon for the first time, four men spent two months living underwater in a subsea laboratory, Tektite I. All at once, a lot of dreams were becoming reality, giving credence to visionary James Lovelock's remark in his book *Gaia,* "We should remember that, in our lifetime, yesterday's science fiction has almost daily become factual history."

The following year, during the Tektite II project, I had a chance to see for myself what it was like to eat and sleep in a warm, dry place 15 meters (50 feet) underwater, then to step through a round hole in the floor and out into the ocean whenever I wanted to go for a swim, day or night. Becoming a resident of the reef with time to stay, observe, and get to know individual fish and their place in the community profoundly affected my understanding of the subtleties—and complexities—that I had missed during in-and-out diving.

The importance of having extended time and of personally being where the action is—in the sea—came clearly into focus for me during this time, but there are depth limitations for humans as divers. Some commercial divers work at depths greater than 300 meters (1,000 feet) using mixes of oxygen, helium, and a touch of nitrogen, and a few have taken brief experimental dives to more than 610 meters (2,000 feet). That means that most of the ocean is beyond diving range—unless you use a submarine.

GOING DEEPER

The advantages of travel in submarines were obvious to some as early as the mid-1600s, when Englishman John Wilkins wrote:

- 'Tis private; a man may thus go . . . without being discovered . . .
- 'Tis safe; from the uncertainties of *Tides,* and the violence of *Tempests* . . .
- It may be of very great advantage against . . . enemies, who by this means may be undermined in the water and blown up . . .
- It may be of unspeakable benefit for submarine experiments and discoveries. . . .

Submarines were prized mostly for military applications until the early 1930s when William Beebe teamed up with submarine designer and engineer Otis Barton to develop a tethered system, the bathysphere, that ultimately took them for a series of dives to as much as half a mile underwater near Bermuda. Crouched within the cramped, 7.6-centimeter-thick (3-inch), 0.9-meter-

Aquarius, *America's space station in the sea in Key Largo, Florida.*

diameter (3-foot) steel sphere that protected them from pressure, Beebe peered through a tiny porthole and witnessed what no human had ever seen before—glowing jellies, fish with lights down their sides like miniature ocean liners, squid that emitted brilliant bursts of luminous ink. In his descriptions of the dives in a 1934 book, *Half Mile Down,* Beebe wrote:

> The only other place comparable to these marvelous nether regions, must surely be naked space itself, out far beyond atmosphere, between the stars . . . where the blackness of space, the shining planets, comets, suns, and stars must really be closely akin to the world of life as it appears to the eyes of an awed human being, in the open ocean, one half mile down.

As a child, I was mesmerized by Beebe's vivid descriptions of life in the deep sea; the reality, it turns out, is even better than the book.

The needle on the depth gauge touched 1,600 feet (490 meters) as I settled the little *Deep Worker* submersible onto the sloping seafloor west of the Dry Tortugas, off Florida. Designed for one person, the sub is more like a diving suit than a submarine. On the way down, I had picked up a following—more than a hundred Caribbean reef squid jetting in tight formation just behind my shoulders, a living cape that turned when I turned, kept pace when I moved forward, held steady when I paused to photograph a fish or crab. Crowds of fast-moving crustaceans flicked around the sub's lights, attracting a number of small lantern fish, finger-size creatures that appeared to be wrapped in foil, sides speckled with blue-green light emanating from what looked like luminous buttons. Occasionally, a squid would break away from the pack and make off with one of the little fish, showering the sub with silver-blue scales.

Rather than serve as a big, glowing fishing lure for the squid, I turned off the sub's lights and sat in the dark, savoring the moment, knowing that no other human had occupied this particular place on the planet, trying to imagine what it must be like to *be* a lantern fish or squid or one of the slender eels that I had seen disappearing headfirst into a hole in the muddy sand and instantly reappearing—headfirst! Our species is fascinated by the poetic notion that somehow, we should all be "reaching for the light," but in the velvety blackness I was more keenly aware than ever that most animals have senses exquisitely tuned to living in the dark.

It is one thing to know that all of life in the sea lives in darkness some of the time, and that most of it never experiences the touch of sunlight, ever. It is something else to sit several hundred feet deeper in the sea than the Empire State Building is high, immersed in liquid darkness, sparked with the faint flash and glow of luminous creatures that give the deep sea the aura of a starry night. Computers, sensors, and cameras on an underwater vehicle can document the terrain, count the fish, take samples, and otherwise do fine surveys, but a robot cannot tell you how it feels to be there, why it wants to go, cannot follow hunches, draw on experiences that cause a pilot to veer away from a given course, laugh, or dream about what to do on the next dive.

When I switched the lights back on, a crab with red eyes and long, feathery legs tiptoed by as gracefully as a Cirque du Soleil performer, just as a giant isopod, *Bathynomus,* lumbered into view. These shoe-size crustaceans, cousins of the tiny "pill bugs" that thrive in my garden, descended relatively unchanged from contemporaries of dinosaurs but were not seen by humans until the 1800s when one was dragged out of the Gulf of Mexico on a fishing line. Now I sat, warm and dry, protected from the nearly four tons of pressure on every square inch that was normal for giant isopods, lethal for me had my body not been encased in a steel cylinder, a protective clear acrylic dome covering my head.

I moved the sub slowly toward a small crimson fish, half-buried in the soft, white sand, becoming so engrossed in filming the little creature's broad, toothy face that I almost missed seeing a flash of silver sweep by at the edge of the sub's small circle of light. An ocean sunfish, *Mola mola,* nearly as large as the sub, glided past—then returned for another look. I abandoned the little red fish, swinging the camera up just in time to record the face and great, horselike eye of the big, disk-shaped fish tilting in my direction, meeting my eyes as it inspected the strange new light-emitting thing that had appeared on the ocean floor. The fish circled several times, then started to move upward, and I followed, ascending until I lost sight of its pale shape about a hundred feet from the bottom. The encounter lasted less than four minutes but left me with a permanent reminder that humans are not the only inquisitive animals on the planet.

Mola mola expert Tierney Thys later confirmed that seeing an "ocean sunfish" (so called because of its habit of basking like a big saucer at the ocean's surface) almost 500 meters (1,640 feet) down was extremely rare—but then, it is extremely rare for people to be looking for a *Mola mola* at that depth. The action I had witnessed would almost certainly have been missed by the narrow field of view provided by the lens of a camera alone.

To be where I was in the little *Deep Worker* in 2001, cloaked in 20th- and 21st-century technologies, required not only that the ingenious engineers who had crafted the sub had gotten their sums right, but also that support had been forthcoming from institutions and individuals who believe that exploring the deep sea must be a priority for our species. In 1997, an invitation from the National Geographic Society to sign on as Explorer-in-Residence coincided with a five-year commitment from the Richard and Rhoda Goldman Foundation to support a project of "ocean exploration, research, education, and conservation" that morphed into the 1998–2003 Sustainable Seas Expeditions, the project that landed me underwater musing about giant isopods, squid, lantern fish, *Mola mola,* their relevance to humankind, and ours to them.

NOAA partnered in the expeditions, providing vital ship support, staff, and expertise in an effort ultimately involving more than a hundred institutions, state and federal agencies, private companies, universities, scientists, teachers, students, and the general public in a private-public, industry-government collaboration that succeeded in establishing baseline data for previously unexplored areas as well as drawing attention to and winning support for the young but promising systems of national marine sanctuaries in the United States, Mexico, and Belize.

For me, one of the most important outcomes of the Sustainable Seas Expeditions was developing heightened awareness for how little is actually known about the ocean, even in areas close to where humans have thrived for centuries. The need to better understand the nature of what was happening to coral reefs, populations of marine life, and the character of the water itself came into sharper focus for many who participated, coincident with a growing global awareness of how much the ocean affects the way the world works—and how much remains to be explored.

RISKS? WHAT RISKS?

Dramatic use of Russia's subs, *Mir 1* and *Mir 2,* occurred in August 2007, an event noted on my cell phone in a message from the sub's chief designer and lead pilot, Anatoly Sagalevitch.

"Sylvia!" the message began. "This is Anatoly calling you from the North Pole. I have just returned from 4,200 meters [13,800 feet] in the *Mir* submersible." There was a pause, and in the background, the sound of glasses clinking, and deep, rumbling laughter. Unmistakably, a celebration was under way. For a decade, Anatoly and I and several intrepid explorers—Don Walsh, pilot of the *Trieste* to the ocean's deepest place, the Marianas Trench; Fred McLaren, captain of the U.S. Navy's nuclear submarine U.S.S. *Queenfish,* during numerous Arctic under-ice excursions; and irrepressible Australian explorer and entrepreneur Mike McDowell—had discussed what it would take to be able to go to the "real" North Pole—4,500 meters (15,000 feet) under the floating ice where Robert Peary and Matthew Henson had stood for the first time in 1909. Several plans were devised, then each was scrapped for one reason or another. McDowell was among the six who finally made the successful descent in the two *Mir* subs.

"We missed you," Sagalevitch continued. "This was a great achievement for science." (More laughter, more clinking of glasses.) "But it was *our* achievement!"

During the expedition, a titanium Russian flag was planted on the seafloor in the heart of the Arctic, thereby establishing Russia's edge on access to the deep sea, and strengthening its claim to the lands under polar ice. Without a doubt, the effect would not have been the same had a robot made the trip.

Several teams are working toward development of manned submersibles to take one or more observers to full ocean depth, the rationale being, "If you can go to the deepest places, you can go anyplace in the ocean." There might be constraints owing to high temperature—as in diving into a subsea volcano or traveling far beneath the ice in polar seas, or getting into narrow crevices, caves, or cracks in the ocean floor, but depth, at least, will not be a limiting factor in ocean exploration when these systems move from vision to reality.

There is a long-standing debate about whether people *belong* in "hostile environments": underwater, crossing the Antarctic continent, exploring caves, or orbiting Earth. NASA Administrator Sean O'Keefe hosted a conference in 2004 to consider questions of risk in exploration, inspired in part by the tragic loss of the *Columbia* spacecraft and its seven crew members. Questions arose about the wisdom of continuing to put astronauts into the sky. Many felt then—and many still do—that a well-crafted machine can more than adequately replace on-site humans almost anywhere, doing almost anything.

In his opening remarks, O'Keefe said: "Over the course of human history every major advance has occurred because of the temerity on the part of human beings to want to understand and to explore and to do something that has not been tried . . . every advance in the course of our existence . . . has been attributed to that characteristic of us as human beings." At another point he admitted, "Yes, risk-taking is inherently failure-prone. Otherwise it would be called 'sure-thing-taking.' "

Most of those in attendance voiced strong support for having people go where it was realistically possible, despite inevitable hazards. Among them was deep-cave explorer William Stone, who said there are two places that he would prefer using robots rather than humans for exploration: in places lethal to humans and in places people can't get to. Clarifying the nature of exploration, he quoted Arctic explorer Vilhjalmur Stefansson, once accused of being an "adventurer." His response was, "An adventure is what happens when exploration goes wrong."

When Charles and Anne Morrow Lindbergh were about to embark on the first east-to-west flight over the North Pole, over territory where no airplane had flown before, a reporter quizzed them about the perils of the journey, their concerns about the risks involved. "Can't you even say you think it is an especially dangerous trip?" he asked. Anne Lindbergh responded, in her book *North to the Orient,* "I'm sorry. I really haven't anything to say. After all, we want to go. What good does it do to talk about danger?"

At the NASA conference, I suggested that "it is true. Danger is the silent partner of exploration . . . but the greater danger is in *not* exploring."

SEND IN THE ROBOTS

New technologies are helping to overcome human limitations—acoustic probes, satellite surveillance, sensors, lasers, innovative diving systems, remotely operated vehicles, autonomous unmanned and manned submersibles, and computers to store, analyze, and communicate the vast amount of data acquired. But the new millennium began with few manned systems capable of accessing even

The author in a Deep Worker *submersible on Florida's Pulley Ridge.*

the average depth of the sea, 4 kilometers (2.5 miles)—the depth where *Titanic* rests on the seafloor. One of these, Woods Hole Oceanographic Institution's deep-diving sub *Alvin,* has logged more successful dives in more parts of the sea than any other submersible. It is one vehicle in an astonishingly small global fleet of deep-diving subs, a workhorse since 1964 that will soon be replaced with a spiffy new, deeper version. Only four manned submersibles in the world, none of them operated by the United States, are currently capable of descending (and returning from) six kilometers (four miles) down—little more than halfway to the deepest places. (Engineers are fond of saying that only round-trips count.)

Meanwhile, since the 1980s, development of remotely operated vehicles (ROVs) and unmanned autonomous underwater vehicles (AUVs) has progressed from relatively simple "flying eyeball" systems to a wide range of small, medium, and some very large systems devised for many jobs previously accomplished only by divers, and many that divers simply could not achieve. The oil and gas industry, in particular, has invested in fleets of high-tech remotely operated systems that inspect, maintain, and repair everything from pipelines to well heads. One even succeeded in gently returning to the depths a large conger eel that had unwittingly been brought to the surface, entangled in a blow-out preventer retrieved for repair on a Kerr-McGee rig in the Gulf of Mexico.

A unique system being built by California-based DOER Marine Operations for Antarctic researcher Ross Powell is 6.7 meters (22 feet) long and will slide down a 59-centimeter-diameter (22-inch) hole bored through 1 kilometer (0.6 mile) of ice, connected to surface controls by a slender fiber-optic line. Once under the ice, the system will unfold like a transformer toy, travel for a week for more than a kilometer, gather samples, and send real-time data back to Powell and his colleagues who would prefer not to be where the vehicle will be. Similarly, robots can now roll over the surface of Mars, while people (some, anyway) are still dreaming of being there themselves.

The Monterey Bay Aquarium Research Institute has fostered development of several large, tethered ROVs that have for years systematically surveyed the region around Monterey Canyon, documenting the nature of life and the overall environment in more detail than exists for most terrestrial environments.

On May 31, 2009, the Woods Hole Oceanographic Institution succeeded in deploying its latest innovation for ocean exploration—the *Nereus*—into the deepest part of the sea, the Mariana Trench. It was the first vehicle to go there since Japan's tethered system, *Kaiko,* made the last of several dives into the trench in 1998. Lost in a storm in 2003, *Kaiko* had been the only system to explore the deepest part of the ocean since the 1960 *Trieste* expedition.

Equipped with a manipulator arm, cameras, and sensors, and on-board batteries for power, *Nereus* can either be operated by pilots on a surface ship who control the vehicle though a lightweight, hair-thin, fiber-optic tether that also sends video images to a large monitor, or it can be switched into a free-swimming, autonomous mode capable of working, and eventually ascending on its own to the surface. Compared with most large ROVs and manned subs, the vehicle is petite, 2 meters (8 feet) wide and 4 meters (14 feet) long, and weighing in at about three tons.

The secret to the success of the system is that *Nereus* is a hybrid, combining attributes that make it possible for the vehicle to fly like an aircraft to survey and map broad areas, then, on the spot, be converted into a pilot-operated system, connected by the thinnest tether yet deployed on an ultra-deep diving vehicle. Most tethered systems use steel-reinforced cables laced around copper wires and optical fibers that transmit power and send information to and from the vehicle. The weight of the tether typically far exceeds that of the vehicle, a tail-wagging-the-dog setup that usually requires sophisticated and costly tether management systems.

The fiber-optic technology used for *Nereus,* largely developed by the U.S. Navy, has opened new possibilities for working in previously inaccessible areas. Andy Bowen, the project manager and principal developer of the system, commented when it was deployed successfully, "With a robot like *Nereus,* we can now explore virtually anywhere in the ocean. . . . I believe it marks the start of a new era in ocean exploration."

MAPPING THE DEEP

With detailed maps of the moon, Mars, and Jupiter in hand, it seems reasonable that the surface of Earth would be known with comparable precision. Not so! Most of Earth's surface is under the sea, of course, and therefore impossible to photograph and map using traditional cartographic techniques. Given the difficulties, it is amazing that it is now possible to envision Earth's major mountain formations—64,000 kilometers (40,000 miles) of mountain ranges running down the length of the major ocean basins like giant backbones—even though few of them have been seen directly. Acoustic technologies make it possible to "see" with sound, and thus chart thousands of peaks and valleys, hills and plains, even when they are covered by several miles of ocean; but the technique has required the use of large ships with appropriate sonar systems "mowing the lawn" back and forth to gather the data needed.

Data from satellites have greatly improved overall knowledge about the configuration of the seafloor far below, and have helped fill in many of the unknowns that existed until the latter part of the 20th century. In National Geographic's *Ocean: An Illustrated Atlas,* Rear Adm. Timothy McGee provides an overview of new technologies that provide hope that better maps of the ocean will soon be possible. Autonomous battery-powered ocean gliders are now assisting in ocean surveys, providing trillions of bits of data per survey—a significant improvement over the single data points that have been normal previously. More data requires more computational power, and this, too, is under way.

Meanwhile, plans are afoot to develop a worldwide network of monitoring stations, a global Integrated Ocean Observation System, that will eventually make possible improved weather forecasting as well as greatly extend and expand knowledge about the ever changing nature of the blue part of the planet.

Knowledge that humankind does have the capacity to alter the nature of the sea may be the most important discovery made so far about the ocean. But the greatest discoveries, of learning how to live within our planetary means, may await. Thanks to generations of curious, daring, intrepid explorers of the past, we may know enough, soon enough, to chart safe passage for ourselves far into the future.

Ancestors of these tube sponges occupied the sea half a billion years ago.

GOVERNING THE OCEAN

The era we are entering will be one of enormous social, political, and economic change—driven in large part from above, from the sky, from Mother Nature. If we want things to stay as they are—that is, if we want to maintain our technological, economic, and moral leadership and a habitable planet, rich with flora and fauna, leopards and lions, and human communities that can grow in a sustainable way—things will have to change around here, and fast.

—Thomas L. Friedman, *Hot, Flat and Crowded,* 2008

A message to graduates from Woody Allen published in the *New York Times* in 1979 began: "More than at any other time in history, mankind faces a crossroads. One path leads to despair and utter hopelessness. The other, to total extinction. Let us pray we have the wisdom to choose correctly."

But there is another path, one that can lead to an enduring place for humankind within the natural systems that sustain us, but only if we understand that the rest of the living world matters. Then, armed with that knowledge, we can actually do what it takes to halt our destructive ways, and going forward, treat nature with proper respect.

To hear more about plans concerning that other path, in June 2009 I attended the 80th birthday celebration and conference in Geneva, Switzerland, of Maurice Strong, a consummate businessman who has helped instigate and implement many of the most effective conservation efforts in history. About 100 guests gathered at the headquarters for the International Union for Conservation of Nature, an institution founded at a meeting in France in 1948 that brought together 18 governments, 7 international organizations, and 107 conservation groups. Strong has had a leading role in fostering the development of the organization that now represents more than 1,000 government and nongovernmental organizations and nearly 11,000 volunteer scientists in more than 160 countries.

As secretary-general of the first major United Nations Conference on the Human Environment in Stockholm, Strong commissioned a report: *Only One Earth: The Care and Maintenance of a Small Planet,* a summary of the findings of 152 leading experts from 58 countries in preparation for the first UN meeting on the environment in 1972. This was a pivotal moment for the relationship of humankind to the environment.

At the Geneva meeting, I listened to well-wishers recall how President Theodore Roosevelt had attempted exactly 100 years ago to convene an International Conservation Conference to work out policies for "world resources and their inventory, conservation, and wise utilization," but plans were canceled by Roosevelt's successor, William Howard Taft. Another broad-based initiative failed at the end of the 1920s when the League of Nations was unable to agree on a proposed world fishery policy, but at least the ideas were planted that eventually made their way into subsequent Law of the Sea deliberations.

Success in gaining formal acknowledgment that the "environment matters" finally came at the 1972 Stockholm Conference, and with it a new kind of multilateral diplomacy—"environmental diplomacy." Among the outcomes was the formation of the United Nations Environment Programme (UNEP), with headquarters in Nairobi. As the agency's first director, Strong convened international experts to consider the implications of climate change, the first such conference on this critical topic.

Despite growing awareness of the truth in former Colorado senator Timothy Wirth's observation that "the economy is a wholly owned subsidiary of the environment," over the 20 years between 1972 and 1992, consumption of the world's natural resources along with planetary pollution accelerated at unprecedented speed, coupled closely with a near doubling of global population. It was the price, some believed, of progress, otherwise termed development. Conflicts over use of land, air, water, wildlife, and minerals provoked numerous new environmental policies and laws, nationally and internationally. In the U.S., the 1970s gave rise to legislation focusing on clean air, clean water, endangered species, wetlands, coastal zone management, protection of marine mammals, and marine sanctuaries.

Late in the 20th century, the ethic of caring for the environment gained momentum, providing hope that there may be ways around Woody Allen's predicted dead ends. "Finding the balance" was the underlying goal of the UN Conference on Environment and Development—best known as the Earth Summit—in Rio de Janeiro, Brazil, in 1992. As secretary-general of the conference, Strong led efforts to find agreement on conventions on climate change and biodiversity. A plan of action known as Agenda 21 was crafted.

I attended the Rio conference as chief scientist of the National Oceanic and Atmospheric Administration, and thanks to Strong, was provided with a pass to attend many of the critical meetings. During one, I watched tensions build as Cuba's President Fidel Castro, notorious for delivering four-hour speeches, stepped to the podium and crammed the equivalent into the precise seven

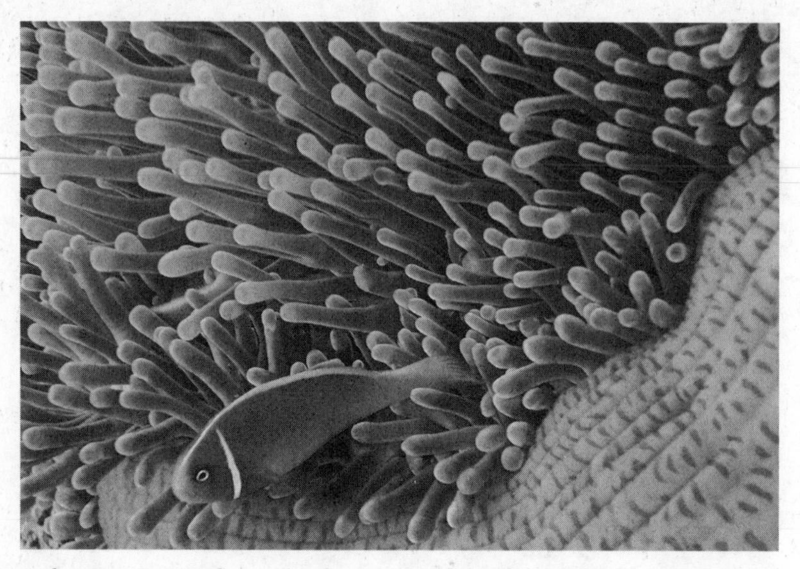

A lasting partnership exists between a clownfish and its anemone host.

minutes allotted to each head of state who spoke. Representatives of 172 governments, including 108 heads of state, attended—the largest gathering of world leaders ever. The speeches, including the one delivered by President George H. W. Bush, acknowledged the connection between a sound environment and a prosperous future for humankind. There emerged an understanding that the economy and the environment are not polar opposites; rather, they are mutually dependent. A sound economy requires a sound environment; a sound environment requires a sound economy. In places where either is in trouble, trouble follows for both.

Agenda 21 acknowledged ocean issues, especially actions relating to pollution and coastal zone management, but major topics were neglected, including recognition of the fundamental role of the ocean as Earth's life-support system, and the urgent need to address protective measures.

The World Summit on Sustainable Development (WSSD), or Earth Summit 2002, in Johannesburg, South Africa, aimed to advance the goals established at the Rio conference ten years earlier. Restoring the world's depleted fish populations by 2015 was one of the ambitious themes agreed to by the world leaders who attended.

By 2012, when the next world summit for the environment will occur, it will be clear whether this and other lofty visions are moving in the right direction—or not.

Well-wishers at the celebration of Strong's decades of "making a difference" noted that in 1972, it seemed there was time to wait for more scientific clarification, time to work things out concerning pressures on the natural world. At the Rio conference in 1992, a sense of urgency for action had grown, and by 2002, in Johannesburg, the concerns had become acute. But now, Strong said, eight decades of experience weighing on every word, "We are out of time. We know what to do. Now we must act."

FREEDOM OF THE SEAS

Nearly every square inch of the world's land has been claimed by one or more nations, as well as the institutions and individuals therein, and reinforced by laws with "dos" and "don'ts," and by policies and claims governing who can do what and where. Even the air has become tightly regulated over much of the planet. But the ocean has had different treatment. Until far into the 20th century, and even now, once away from the shore, the open ocean is regarded as a place where anyone can enjoy a certain "freedom of the seas." The concept technically provides freedom of access for ships of any nation to travel unhampered through international waters, and ensures the right of neutral shipping for trade in wartime except where blockades are established. But it also has generally meant the freedom to take what is there, from fish to goods lost in the deep sea, and to dispose of unwanted and noxious things.

The notion of "ownership," whether land, air, or sea, is relatively new in the history of humankind, an inevitable consequence, perhaps, of population growth and increasing pressures on the use of shared space. It seems a very human thing to do, to own and defend territory and property, but we are not unique in this tendency. Male mockingbirds declare to all within earshot that a certain patch of trees and shrubs is theirs; giant jawfish challenge other giant jawfish to keep a respectful distance from staked-out turf in the bottom of the Gulf of California or suffer fierce mouth-to-mouth combat; even ants go to war over real estate. Had they been consulted, the gray whales that migrate thousands of miles to four unique lagoons along the coast of Mexico's Baja peninsula would most likely have objected to plans by a Japanese-Mexican consortium to construct a large salt-processing plant at San Ignacio, "their" breeding and nursery area.

Don Marquis described this not entirely human attribute in his ode to "warty bliggens, the toad," in his 1927 classic *archy and mehitabel:*

i met a toad
the other day by the name
of warty bliggens
he was sitting under
a toadstool
feeling contented
he explained that when the cosmos
was created
that toadstool was especially
planned for his personal
shelter from sun and rain . . .
a little more
conversation revealed

that warty bliggens
considers himself to be
the center of the same
universe
the earth exists
to grow toadstools for him
to sit under
the sun to give him light
by day and the moon
and wheeling constellations
to make beautiful
the night for the sake of
warty bliggens . . .

i asked him
why is it that you
are so greatly favored

ask rather
said warty bliggens
what the universe
has done to deserve me . . .

Like warty, humans tend to believe that we are entitled to all of whatever there is, no matter what the mockingbirds, jawfish, ants, whales, tunas, toads, and millions of other forms of life with whom we share space might think.

While we may not be alone in believing that the world, or even some part of it, "belongs" to us, no other species seems to have a comparable sense of awareness of the consequences of actions that could work against our ultimate survival—and to care not only about the present, but also future generations, and even the fate of other forms of life. Maybe it is a deeply embedded survival streak; or it may have overtones of altruism, another feature that is manifest from time to time in other species. Whatever it is that gives us something of an edge as a species, we are just beginning to realize the magnitude of the destructive impacts we have had, and continue to have, on the natural world that sustains us, and to be concerned about the implications—for us. This is especially true of our relationship to the sea. Now that we know, it is urgently important to rethink policies that have caused us to reach the perilous place where we have now arrived.

The "Wild West/anything goes" attitude that once characterized all of the ocean has been reined in loosely, although the "high seas"—the 64 percent of the ocean beyond waters claimed by various countries—still functions as a place where individuals, corporations, and countries can travel without constraint, and for 11 nations engaged in industrial fishing, it is largely a great blue free-for-all.

In the 1400s, there were attempts by Spain and Portugal to claim shipping rights over large parts of the ocean, both believing that the seas were there to serve them—a bold, warty bliggens kind of idea that had little hope of being enforceable. The inability to control or defend claims to the ocean gave rise in the 17th century to the idea of *Mare Liberum,* literally, "freedom of the sea," by the Dutchman Hugo Grotius. A corollary of the concept was the presumption that the ocean was an inexhaustible resource and did not require restrictions. Until the late 1700s, nations did not claim effective authority over any part of the ocean, except in areas where societies defended their "rights" to local populations of fish and other valued resources.

Since World War II, numerous international treaties have been developed through the United Nations and subsidiary institutions relating to slavery, piracy, drug trafficking at sea, and conservation issues. The list is long, but there are now hundreds of national, regional, and global environmental regulations relating to fisheries management, marine mammals, waste disposal, oil spill response, and climate change.

LAW OF THE SEA—AND BEYOND

Currently, the primary mechanism for international policy and governance over the behavior of nations with respect to the ocean is the United Nations Convention on the Law of the Sea. Drafted in the 1970s, it was opened for signature in 1982, signed by 119 countries that year, and ratified by the requisite 60 nations needed to bring the treaty into force in 1994. It is now binding on all of the 55 nations, plus the European Union countries, that have ratified it. The United States is the only major maritime power that, despite its key role in framing the Law of the Sea, is still not officially a party to it. President Bill Clinton signed for the United States in 1998, but more than a decade later, the U.S. Senate continues to resist ratification.

Nonetheless, as the 21st century gets under way, Law of the Sea provides a comprehensive global legal framework that governs human activities on and in the world's ocean. Hundreds of pages of fine print, painstakingly hammered out during years of intense negotiations over words, paragraphs, principles, and even punctuation, define the rights of military mobility on the high seas, through international straits, and in coastal waters; the free movement of global commerce; high-seas freedoms for laying cables and pipelines; an international framework for maritime law enforcement; marine environmental protection; marine scientific research; and creation of a mechanism for settling international disputes.

The provision of the convention that allows coastal states to declare jurisdiction over living and nonliving resources in a zone up to 200 nautical miles off their shorelines, the Exclusive

Economic Zone (EEZ), heightened the interest in the ocean for many countries, especially small island nations, which suddenly had a much larger presence in the world.

The United States independently claimed an EEZ in 1983, more than doubling the area of the planet under the nation's jurisdiction, but not until the national Oceans Act of 2000 was a comprehensive review of U.S. policy concerning the sea called for—the first since 1960. A National Ocean Commission was appointed soon after an independent, nongovernmental U.S. initiative—the Pew Oceans Commission—was formed.

The Pew report, entitled *America's Living Ocean: Charting a Course for Sea Change,* mirrored many of the findings of the national commission, and in due course, the two came together in a Joint Ocean Commission initiative.

In 2002, while the two U.S. commissions were deliberating U.S. ocean policy, I proposed to Intel co-founder and conservationist Gordon Moore that it would be timely to convene a conference to assess *international* ocean issues, one that might have an impact comparable to an event organized by Conservation International at the California Institute of Technology in 2000, called Defying Nature's End (DNE). The scientists, economists, and business leaders at the DNE conference were charged with defining actions that could stem the loss of global biodiversity, with an emphasis on "hot spots"—areas of high diversity and high threat—as well as intact wilderness areas. The conference action plan turned out to have pivotal impact, spurring actions to demonstrate the economic, societal, and security advantages of having a world rich with healthy, diverse ecosystems, while underlining the problems that arise when they are lost.

Reporting on the conference in 2001, the journal *Science* noted that "greatly increasing the areas where biodiversity is protected is a clear and achievable goal, one potentially attainable by using funds raised in the private sector and leveraged through governments." Critical needs were identified: synthesis and distribution of available knowledge, investment in targeted research, development of research and management centers in countries where biodiversity actions are focused, and demonstration of links among biodiversity, ecosystems, their services, and people—all leading to benefits for humankind through increased and enduring protection for hot spots and wilderness areas worldwide.

Some consideration was given to coastal and ocean matters in the Defying Nature's End conference, but both the recommendations and the implementation focused largely on the land. A follow-up for the ocean seemed timely, and in 2002, the Gordon and Betty Moore Foundation agreed to back a yearlong effort to assemble data to prepare for and hold an event to consider how to make a difference for the ocean. In May 2003, 150 scientists, policymakers, educators, economists, communications professionals, and business leaders from 70 organizations and 20 countries gathered in Los Cabos, Mexico, in May-June 2003 for the "Defying *Ocean's* End" (DOE) conference.

A century ago, English biologist Thomas Huxley articulated the underlying motivation for taking on such an ambitious challenge in his book *Man's Place in Nature:*

> The question of questions for mankind—the problem which underlies all others, and is more deeply interesting than any other—is the ascertainment of the place which man occupies in nature, and of his relations to the universe of things.

The fundamental focus for DOE participants was: What is our place in the sea?

To the consternation of some and to the delight of others, sea creatures were deliberately omitted from the otherwise hearty

Mexican menu provided throughout the weeklong meeting. It did not seem right, somehow, to be discussing how to protect the the creatures of the sea with none of them present at the table—except cooked.

The goal of the conference was to develop a global action plan—with priorities, timetables, and estimated costs—aimed at stabilizing and reversing the devastating decline of ocean systems. Graeme Kelleher, the Australian visionary who headed the Great Barrier Reef National Marine Park Authority for more than two decades, chaired the meeting, and reviewed the following key developments, which provided a foundation for the deliberations:

- The United Nations Convention on the Law of the Sea (UNCLOS) in 1992
- The Convention on Biological Diversity (CBD) in 1992
- The United Nations Millennium Summit in 2000
- The United Nations Conference on the Environment and Development in Johannesburg in 2002
- The World Summit on Sustainable Development (WSSD) in 2002
- The establishment of a High Seas Marine Protected Area (MPA) Executive Committee, which brings together the International Union for Conservation of Nature (IUCN), the World Wildlife Fund (WWF), the World Commission on Protected Areas (WCPA), and some governments in the cause of high-seas MPAs, in 2003

Clearly, there is international interest in trying to figure out what to do with the dangerous depletion of the natural systems that are vital to global economies, health, security, and to life itself. While the first step toward solving problems is knowing that you have them, the next moves are trickier. What to do?

WORKING ON SOLUTIONS

At the DOE conference, several areas were chosen to serve as models for action plans that could help locally and regionally: seamounts in the high seas, the Southern Ocean, the Patagonian shelf, the Coral Triangle in the tropical Pacific Ocean, the Gulf of California, and the Caribbean Sea. Success in any of these areas could serve to inspire action elsewhere; and taken as a whole, global progress could be realized.

The area that brought most of the issues—governance, fisheries, ocean use planning, exploration, technology, and communication—most clearly into focus is the region known as the broader Caribbean, which includes the Gulf of Mexico, the Bahamas, and Turks and Caicos.

The region encompasses the heart of biodiversity in the Atlantic Ocean, but within a relatively small geographical area, it features remarkable sociopolitical complexity. Human pressures have altered the environment there for millennia, but as the DOE team that reviewed the region confirmed, "There is no doubt that current pressures are unparalleled."

Following a familiar pattern, archaeological evidence shows that extinction of many terrestrial species followed the arrival of people, even when populations were low. In the late 1400s when European explorers came to the Caribbean, they encountered two dominant groups of people living among the islands, as well as smaller populations of other lineage. Large animals—manatees, monk seals, and turtles—had already declined as a consequence of hunting, and it appears that subsequent loss of indigenous peoples enabled a temporary recovery of the creatures they had traditionally taken. That soon changed, however, as new arrivals from Europe, Africa, and North America supplemented the crops they cultivated with catches of whales, seals, turtles, conchs, lobsters, and many kinds of fish. From 1688 to 1730, a dedicated fleet took about 13,000 turtles each year from Grand Cayman Island to help feed a large British colony in Jamaica; by the end of the 18th century, the Grand Cayman turtle population had essentially been eliminated. Toward the end of the 1800s, the taking of turtles, mostly green sea turtles for export to England, spread across the Caribbean, with predictable consequences.

The latter half of the 20th century was marked by an economic shift from reliance on agriculture and fishing to a strong dependence on tourism, with dive tourism alone producing more than a billion dollars a year by 2005; but fishing did not stop. Rather, pressures increased with the ability to serve new markets and use new technologies to access previously remote areas. Bottom trawling for shrimp and "ground fish" not only dramatically altered the populations of numerous species but also destroyed large areas of the ocean floor, especially in places that were scraped repeatedly.

Human population in the Caribbean region has more than doubled from the 1970s to the early part of the 21st century, with numbers now close to 50 million. The coincident problems for the ocean are easy to find: upstream sources of pollution, coastal development, loss of nearshore habitats, decline of critically important fish and other species targeted for consumption—the list is long and dreary. More challenging are the solutions.

The DOE team analyzed the numerous environmental agreements and protocols relevant to the Caribbean, notably the Cartagena Convention—the Convention for the Protection and Development of the Marine Environment of the Wider Caribbean Region, an example of one of the more "advanced" regional seas conventions that builds on the Law of the Sea. While much remains to be explored and discovered about the nature of the ocean systems and creatures that live in the region, it stands out as one of the most researched areas of the world—enough, the DOE report concluded, "to challenge the trends of anthropogenic (human-made) impacts."

Schooling fish form moving communities with fluid boundaries.

When the yearlong analysis of the various areas and issues was assembled, a lively week of intense meetings and two more years of follow-up discussions led to these key recommendations, published in 2005:

- Treat the 60 percent of the world ocean outside of national EEZs as a World Ocean Public Trust, with legal approaches concerning use of the high seas, including fisheries, under coordinated, international, multi-use zoning regimes.
- Reform fisheries using market-based mechanisms, subsidy changes, and sustainable practices.
- Implement global and regional communications plans to educate the public.
- Create, consolidate, and strengthen marine protected areas into a globally representative network.

- Develop an expanded research program focused on top priority marine environments high in endemism and biodiversity.

Events following the DOE conference have underscored the urgency of taking action. Among them are:

- The World Parks Congress in Durban, South Africa, in 2003
- The seventh meeting of the Conference of the Parties (COP 7) to the Convention on Biological Diversity in Kuala Lumpur, Malaysia, in 2004
- The third IUCN World Conservation Congress in Bangkok in 2004, followed by the fourth in Barcelona, Spain, in 2008
- The first International Marine Protected Areas Congress in Geelong, Australia, in 2005, followed by the second in Washington, D.C., in 2009
- The World Ocean Conference in Manado, Indonesia, in 2009

Around the world, people are responding to the growing knowledge that our place on the planet is at risk, and there are no ready options about another place in the universe to go.

OCEAN CARE: A MORAL ISSUE?

At the DOE conference, Chair Graeme Kelleher noted that some have interpreted the biblical exhortation "to rule over the fishes of the sea, and the birds of the air, and all living creatures that move upon the earth" as a mandate, an obligation of sorts, to consume the natural world. But, he said, "If there is one lesson that the problems of our environment should have taught us, it is that this injunction is disastrous." He then asked those assembled to deliberate solutions to this problem. "Stewardship" is a slightly more palatable way of expressing our place in the greater scheme of things; but it still smacks of the warty bliggens syndrome.

Patriarch Bartholomew I, spiritual leader of the world's 250 million Orthodox Christians, has declared, "For humans to cause species to become extinct and to destroy the biological diversity of God's creation, for humans to degrade the integrity of the earth by causing changes in its climate, by stripping the earth of its natural forests, or destroying its wetlands, for humans to contaminate the earth's waters, its land, its air and its life with poisonous substances, these are sins."

Pope John Paul II affirmed that "the ecological crisis is a moral issue." Whatever the underlying morality, the fact is that our actions relative to the natural living world, land, air, and sea, have taken us to a precipice, a tipping point, a crossroads with ourselves in the crosshairs, the ones responsible for the fix we're in. We cannot magically go back to a time when the carbon dioxide level in the atmosphere was less, when fish, trees, and wild places were more, but we can make the future better than it otherwise would be if we do nothing. Time is running out for choices that we can act on now. If we do not, we will forever lose the chance.

Carl Safina asks in his *Song for the Blue Ocean:*

> Which will it be, then: degradation or recovery, scarcity or plenty, compassion or greed, love or fear, ahead to better times or to worse? We will all, by our actions or inaction, help decide.

Cultivating carnivores such as these barracuda is unrealistic.

SMART AQUACULTURE

10

When we take fish from the ocean, we are not farmers harvesting the fields we have sown and tended; rather we are hunters, the top predators in the ecosystem, more efficient and more voracious than any other inhabitant of the marine food web.

—Daniel Pauly and Jay Maclean, *In a Perfect Ocean,* 2003

Soochow (Suzhou), China, is an ancient city near Shanghai renowned for magnificent gardens and great marshes extending from a southern branch of the Yangtze River. It is a special place for me because it was there that I had my first look at one of China's famous fishponds. As the token scientist, I had joined a small group of women invited to meet with professional counterparts in 1973, a visit that included meetings with officials in various cities, tours of hospitals, schools, museums, factories—and an elaborate many-course banquet on a commune where everything served was raised or caught within a few miles of where we sat.

After walking through fields of beans, cabbage, garlic, and other vegetables, we were guided to a raised area where closely pruned mulberry trees surrounded a distinctly green but well-kept pond.

A dozen ducks paddled to the far side of the pond, giving way to several men armed with a net stretched between two poles.

One of our group, Alison Stilwell Cameron (daughter of World War II General "Vinegar Joe" Stilwell), had spent her first 16 years in China, and she filled us in on what was about to happen. "They want to show you the fish," she said. "They have five different kinds of carp growing in there, all eating the algae that are making the pond green or the little planktonic animals that also eat algae. The ducks fertilize the ponds and produce eggs that are eaten, with enough saved to produce more ducks—including some that are eaten as well. In China, carp are called 'home fish' because they can be cultivated even in small pools."

"And the trees?" I asked. "Silkworm food," she said. "The caterpillars thrive on mulberry leaves, and the trees thrive on the fish- and duck-enriched pond water. Nothing is wasted."

With seven billion people to feed on a planet that stubbornly remains the same size, we seriously need to avoid waste. Smart aquaculture could help take the pressure off ocean wildlife now being caught for food and commodities, and it might yield large new sources of readily available, high-quality protein. But serious issues need to be to be resolved.

FARMING FOR PROTEIN

Dinner on the commune featured rice, at least one kind of carp, several varieties of beans, cabbage, shredded duck, and a shimmering mound of what at first appeared to be translucent noodles, until I realized that what I thought were speckles of black pepper were the eyes of hundreds of small eels, served to us as an exceptional treat. Each slim creature, captured as it was heading upstream from a distant ocean spawning site, would normally have been raised in the commune's ponds to become a 1-meter-long

(3.3-foot) adult. Dessert—clusters of sweet, golden loquat fruit—was preceded by bowls of water lily soup, dozens of crisp, young, still curled leaves sheathed in jelly floating in a clear broth.

For thousand of years, Chinese farmers have been combining agriculture and aquaculture for an efficient and effective use of land and water. They nourish rice and fish in the same flooded paddies, using wastes from humans and farm animals—pigs, chickens, ducks, geese—to provide nutrition for crops. They have perfected the cultivation of several kinds of hardy freshwater fish that grow fast, eat plants, taste good, reproduce in captivity, and can be raised in large numbers in closed systems. When the goal is to obtain great quantities of high-quality protein with minimum cost to the farmer and to the environment, these are good features to look for.

Eels, unfortunately, do not qualify as good candidates for high-volume cultivation because they spawn in the open sea. There they spend two or three years undergoing an amazing transformation from eggs to transparent juveniles that resemble glass leaves more than they do proper fish, let alone eels. As predators, their diet changes, too, shifting over the years, from minute planktonic animals when they are at sea, to insects, fish, and other prey when they become adults. The eels are still very small when they arrive in swarms at the rivers where their parents once traveled, after decades of living in fresh water miles away. Having baby eels for dinner came at a high cost to the ancient ways of these creatures, animals that for millions of years did not have predatory humans on the long list of hazards to their survival.

A few mammals and birds have proved to be suitable as cultivated conveyers of energy from the sun, turning plants into pounds of protein for human consumption: cows, pigs, sheep, goats, chicken, turkeys, ducks, and a handful of others. All are primarily herbivores, brought to market within a year or so of birth.

They can live in fairly confined spaces and raised through many generations, and all have wide culinary acceptance. There are good reasons why lions, tigers, wolves, and hyenas have not been selected as farm animals. It is hard to imagine a serene enclosure filled with several hundred lions, but even harder to think of the high cost of keeping them in calories.

Most of the calories that power people come from a remarkably small number of plants and animals, considering the millions of possibilities that surround us. If we really wanted to be serious about obtaining food most efficiently to feed large numbers of people, we would be focusing more on tasty microbes, less on raising cows or taking top carnivores out of the sea. Photosynthetic microbes capture the sun's energy and can double their mass in days, taking up carbon dioxide and generating oxygen along the way. This is the ultimate source of energy for most of life on Earth: life in the sea. Why not for us? Of course, chefs would find it challenging to make microbes appear irresistible, but then, oysters are not that appealing at first glance, either.

FARMING LOW ON THE FOOD CHAIN

Some species of blue-green bacteria *are* being raised in closed ponds and tanks and used to feed people and animals, the products appearing in various drinks and capsules. During a National Geographic expedition to the Galápagos, I learned over early morning coffee about a company, Martek Biosciences Corporation, that is successfully using tanks located away from the ocean to cultivate large quantities of the minute marine organisms that produce the omega-3 fatty acid DHA. Fish oils containing omega-3 are popular as a dietary supplement, as well as for use in animal food and various products, but obtaining them requires catching and killing large numbers of menhaden and other oily

Closed systems provide hope for successful aquaculture.

fish. Henry "Pete" Linsert, then CEO of the company, said, "Fish don't make the DHA themselves; they acquire it from plankton. We figured, why not give the fish a break and go straight to the source?"

Growing large varieties of seaweed for human consumption as well as for products is an aquacultural art developed in Asian countries for centuries. At the coastal city of Tsingtao (Qingdao), famous for its fine beer, naval base, and the world-renowned Institute of Oceanology, I had a personal tour of the nearby kelp farm in 1980 from the institute's founder and director, Cheng Kui Tseng. A seaweed specialist with a doctorate from the University of Michigan, Tseng is a hero to some of us for being the first diver at the Scripps Institution of Oceanography in 1943, but many others have profound respect for his pioneering breakthroughs in the cultivation of marine mollusks and algae, including the popular red alga sold as nori in Japan, but known to seaweed specialists as *Porphyra*.

Tseng led me into a large domed greenhouse where bubbling seawater flowed over rows of shallow trays filled with thousands of empty clamshells. "This is where the *Porphyra* gets its start," he explained. The thin, slippery blades grow from spores released from a small filamentous phase growing on the clamshells. This phase of the plant looks nothing like the familiar nori sold in markets, and for many years it was classified as a different kind of alga altogether. So important was solving the mystery of *Porphyra*'s life cycle that a monument was erected in Tokyo to honor Kathleen Drew Baker, the British scientist who figured it out.

"*Porphyra* is just one kind of seaweed we're farming," Tseng told me. "Mostly, we are focusing on the kelp. The *Laminaria* in the bay is a species originally from Japan, but here it grows more quickly, and we have been doing some selection to encourage the fastest-growing and best-quality forms."

Some of the kelp is eaten raw or dried, but much of it is turned into various products. The giant kelp *Macrocystis pyrifera,* which grows along the coast of California, and related species in Peru, Chile, South Africa, and other cool-water areas—as well as certain kinds of fleshy red algae—yield gelatinous colloids that turn up in hundreds of places: soups, sauces, mayonnaise, chocolate milk, ice cream, cheeses, bakery goods, fruit syrups, aspics, puddings, candies, toothpaste, dental impression materials, lotions, paints, plastics, tapes, decals, shipping containers, culture media for microbiologists—the list is long and growing.

At the ninth International Seaweed Symposium at the University of California at Santa Barbara in 1977, Michael Neushul, another pioneering scientist in the realm of seaweed cultivation, observed, "It's time to design a new world food and energy system geared to renewable energy. . . . It is worthwhile to explore the potential of marine plants as collectors of energy and concentrators of nutrients, since these might contribute significantly to a solar-based food and energy production system." More than 30 years later, we have made little progress toward this vision, but the groundwork for eventual success was laid by Neushul and a fellow biologist and engineer from the California Institute of Technology, Wheeler North. In the 1970s, North secured funding to construct an underwater kelp farm, an enormous umbrella-like structure placed offshore near Los Angeles. The farm worked fine in principle but did not prove to be practical for large-scale, commercial production.

LEARNING FROM HISTORY

China continues to lead in freshwater and ocean aquaculture, contributing about half of all cultivated aquatic organisms produced in the world. In these aquatic farms, as elsewhere, the 10,000-year development of agriculture is now compressed into a few decades.

In early forms of agriculture, people "managed" wild plants and animals, giving naturally occurring populations special protection, care, or even modest enhancement, such as providing food or inhibiting predators. Protecting spawning areas or natural oyster beds, or taking sea creatures only in certain seasons, are comparable mild methods of aquaculture management.

The second phase of agriculture builds on the first but goes farther with manipulation of the environment and the species. Examples in aquaculture include providing artificial structures for oysters or open-sea cage cultivation of enclosed species. Coastal shrimp and oyster farms are in this category. So is salmon farming. The life history of the fish is controlled, but its environment is not. This is true even of the futuristic submersible self-propelled current-riding geodesic-sphere fish pen recently launched

Fish farmed in a closed system near Kuching, Sarawak, Malaysia.

in the Caribbean, filled with baby cobia. The pen does not control the ocean environment, and the carnivorous cobia have not been raised from a farmed population. There are still some big loose ends to tie up before aquaculture will have the kind of efficiency attained in terrestrial systems that deliver food to billions of people.

The focus on carnivorous animals works if the goal is to serve relatively small, high-end luxury markets, but if a large volume of quality food is the goal, plants and plant-eaters are the way to go—just as on the land. Mindful of the problem, backers of Kona Blue Water Farms in Hawaii are seeking better kinds of food for the fish they have chosen to grow: the Hawaiian yellowtail, marketed as Kona Kampachi. Originally fed largely on anchovies from Peru, the yellowtail currently eats more soy and fewer fish.

The third kind of aquaculture, the approach most likely to yield enduring benefits, involves true domestication: control of the life cycle of the desired species and of the environment in closed systems. Freshwater cultivation of carp in ponds linked to ducks and trees fits here, as does closed-system aquaculture at the Mote Marine Laboratory in Sarasota, Florida. There, large sheds house tanks cycling water through microbial "scrubbers" that remove nitrates and phosphates and return cleansed water to the fish. "It is like a big aquarium," explained Kumar Mahadevan, director of the lab. "At our public aquarium the system works in a similar way, with water constantly being cleansed and returned."

Advances in managing animals and water in public aquariums in recent years have given closed-system aquaculture a significant boost. Raising tropical fish for aquariums has become big business, and not just for major facilities such as the new Georgia Aquarium in Atlanta, where more than half of its thousands of finny residents have been raised in captivity. Many home aquarists have moved beyond having a few goldfish in a glass bowl and are making breakthroughs in cultivating coral, sea stars, delicate shrimp, and crabs that previously have eluded even determined aquarists.

On the Bahamian island Eleuthera, Chris Maxey, Navy Seal turned headmaster of the Island School, encourages students to learn self-sufficiency coupled with community service. He is experimenting with aquaculture on two levels. The teenagers learn to dive, drive boats, and tend a large net anchored about a mile offshore that holds cobia—a local fish that grows fairly quickly on a diet of fish-filled pellets. Under a canopy on the shoreside campus are large, round tanks, some containing the freshwater fish tilapia, and others, clusters of lettuce growing hydroponically. Water enriched by nitrates from the fish nourishes the lettuce, and then that water, purified and oxygenated, flows back to the fish in a large, looping figure eight.

A WORLD OF CHOICES

Given the swift decline of wild fish at all levels, from plankton-eating anchovies and menhaden to carnivorous squids and sharks, capture fishing can no longer be viewed as a viable way to feed the world today, let alone a growing population of people. Ocean scientist Jeremy Jackson told a March 2009 meeting in Washington, D.C., "To get a lot of protein from the sea in the future, aquaculture is the answer." But it has to be the right kind of aquaculture.

According to a 2000 report from the Food and Agriculture Organization of the United Nations (FAO), about a billion people—one in seven—rely on fish from the sea as their primary source of protein. In North America, about 10 percent of the animal protein consumed comes from fish, wild and farmed, but in China, the level is 22 percent.

To the extent that wild-caught animals contribute to these numbers, trouble is brewing, because fish populations are collapsing. At the present rate of decline, by mid-century the number of people who can rely on wildlife from the sea will be vanishingly small. In theory, aquaculture could fill the growing void, but despite the urgency, solutions are slow in coming.

You would never guess that problems are looming by looking at the seafood spread in my local market in Oakland, California. It looks like the United Nations of Fish, dozens of wild and farmed sea animals in neat rows draped over sparkling mounds of ice:

(All prices are per pound, except oysters)

Ahi tuna, Vietnam	7.99
Ahi tuna, Philippines	15.99
Orange roughy, New Zealand	9.99
Sea bass, New Zealand	19.99
Catfish, farmed, U.S.	5.99
Catfish nuggets, farmed, U.S.	1.99
Tilapia, farmed, China	5.99
Petrale sole, wild, U.S.	11.99
Salmon, farmed, U.S.	1.99
Atlantic salmon, Canada	
(Natural color processed in U.S.)	6.99
Atlantic salmon, color added	4.99
Calamari, Taiwan	5.99
Mussels, farmed, Canada	2.99
Oysters, Pacific, farmed	1.00 each
Large sea scallops, U.S.	9.99
Shrimp, farmed, Thailand	4.99
Shrimp, farm-raised (no country)	5.99
Blue crab, wild, U.S.	12.99
Red king crab legs, Alaska	9.59
Snow crab legs	6.99
Stone crabs, farm-raised	7.99
White shrimp, U.S.	9.99
Lobster tail, Canada	19.99
Lobster tail, Brazil	27.99

It's really confusing. Some doctors say, "Eat seafood, it's really good for you. High-quality protein and valuable omega-3 oils, especially from tuna, swordfish, salmon, and cod."

Other doctors say, "Be careful! Tuna, swordfish, shark, cod, orange roughy, halibut, hoki, sea bass—all have high levels of mercury and other toxic substances. It's especially a concern for expectant mothers, but not really good for anyone, at any age."

"Eat salmon," some advise, "but watch out for the farmed ones. They cost less, but are artificially colored and have been treated with antibiotics. Go for the wild ones!"

"Whatever you do, don't eat the wild salmon," some of my conservation-minded friends caution. "It's OK if you want to catch them yourself, but there simply aren't enough to supply the demand in San Francisco, Miami, Chicago, London, Paris, Sydney, Tokyo, and everywhere else that people hanker for wild salmon. There must be someplace where the eagles and bears don't have to compete with people in New York restaurants for their dinner."

SO, WHAT SHOULD WE EAT?

What are the smart choices? Is any seafood safe to eat? If so, what?

Thoughtfully prepared seafood guides are provided by organizations in many countries, some ranking choices primarily on the level of contaminants, others focusing on three factors for captured fish: state of the population, amount of bycatch, and level of habitat destruction. Failure to get a passing grade for any of these criteria may relegate a species to the "red—don't buy" category. I can easily pass up the entire wild-caught contents of the seafood counter knowing what I do about how important the animals are when alive, doing their part to maintain the systems all of us rely on for our existence; I know how precarious their future has become owing to our insatiable appetite for them. I figure that I can help their chances for survival—and for the survival of my species, too—by choosing not to eat them. As Peter Knights, director of WildAid, says about tigers, elephants, rhinoceroses, sharks, and other endangered wildlife, "When the buying stops, the killing can too."

The various seafood guides usually rank farmed fish based on safety for consumers as well as on environmental impacts. Currently, it's really hard to find out where or how animals were raised, what they have consumed that you don't want to have as a part of you, or how long they have been sitting in storage, accumulating things you also do not want to have as a part of you. What most guides do not tell you is whether the fish are plant-eaters or carnivores, nor do you learn their likely age, and these things matter a lot for two reasons. The higher up the food chain, and the older the animal, the greater the concentration of contaminants: tuna, shark, swordfish, halibut, and in fact, most of the fish in the counter fit into this category. It takes a much greater investment from the ecosystem, pound for pound, to make a ten-year-old fish-eating tuna than a one-year-old plant-eating catfish. For those who want to eat low on the food chain with lowest risk of contaminants, farmed catfish, tilapia, carp, and certain mollusks are the best choices, but even so, it makes a difference where and how they were raised.

About the many kinds of fresh, frozen, peeled or whole, small, medium, and large shrimp? Wild shrimp have been totally off my menu since I first went aboard a shrimp trawler in the 1950s and saw what anyone can witness vicariously by going to see the popular film *Forrest Gump*. The big net is winched aboard, the end of the trawl opens and, *wham!* Onto the deck spills an avalanche of dead and dying animals: young redfish, little flounders, sea trout, rays, urchins, sea stars, sponges, whip corals, crabs, a mass of tortured animals writhing like a scene from a Hieronymus Bosch painting, and here and there, the flicking jump of a shrimp, instinctively doing everything it can to get back to the soothing embrace of the sea.

Now that I know what it takes to bring shrimp to the table, I can no longer face them on a plate. The Monterey Bay Aquarium once had an exhibit, "Fishing for Solutions," that included a case showing six luscious-looking prawns artfully draped over a crystal goblet. Arching over it was a shadowy portrayal of the bycatch—the fish, turtles, sea stars, and other creatures killed in the process of catching the half dozen shrimp that are the real cost of the shrimp cocktail.

Sven Lindblad, explorer and ecotourism pioneer, has decided not to serve shrimp on any of the seven ships that he uses to take visitors to wild and wonderful places on the planet. Many guests expect, and some demand, a menu that includes seafood. Lindblad provides choices, but no shrimp, and explains why not. Wild-caught shrimp? Too many problems with bycatch, habitat destruction, and overexploitation. Farmed shrimp? Some are better than others, but it is difficult to identify the source when buying in large quantities, and most involve large-scale loss of coastal mangroves, marshes, reefs, and other critically important coastal systems. Until current problems are resolved—there is always chocolate.

So what about the mussels? Salmon? Catfish? Tilapia? Carp? Stone crabs?

First the stone crabs. It is hard to imagine how these tasty meat-eaters could be *farmed.* Raising any species of crab from egg to adult is a complicated business, requiring special food for the various microscopic planktonic stages crabs go through before they look like miniature adults. Catching and caging juveniles and feeding them until they grow big enough for market doesn't really count as farming, whether it's crabs, tunas, or lobsters being "fattened" for sale. Unlike typical farm animals, which grow to market size in a year, it takes about three or four years for a stone crab just to grow from an egg to young adulthood. The issue is similar for some big fish: Caging and growing out young bluefin tuna means seriously interfering with the chances

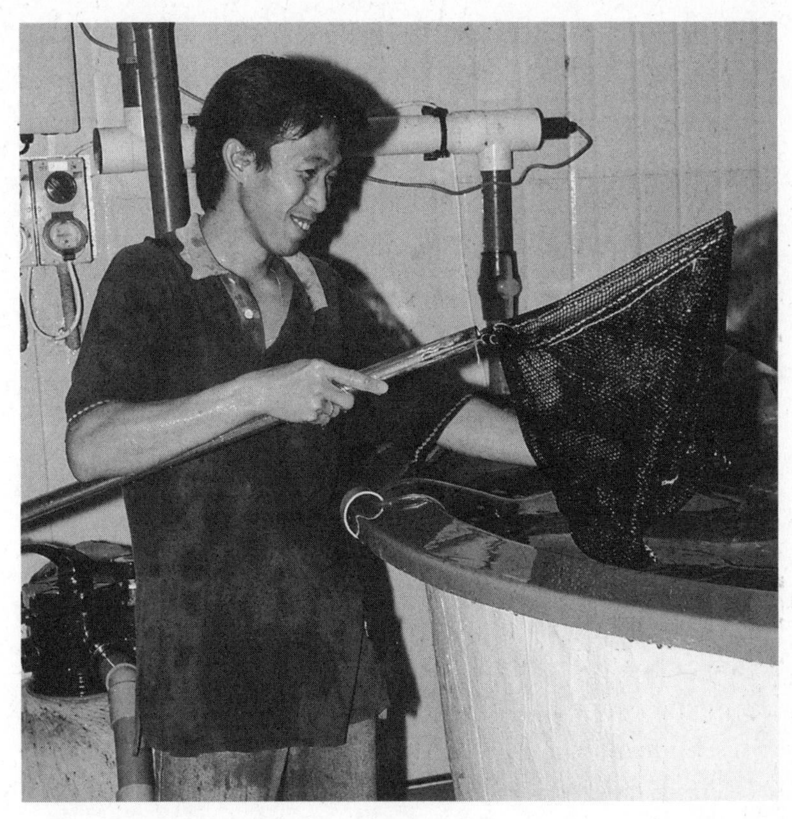

Growing fish in tanks has advantages over open sea cultivation.

for these greatly depleted animals to recover. It's economically viable only because the price for tuna is high enough to make the investment work.

What about mussels, oysters, and clams? If you know the source, know where those filter-feeding mollusks have been vacuuming up their sustenance (which becomes a part of you), well, fine. These mollusks are among the most efficient filter-feeders on the planet, each one able to draw in more than a gallon of water an *hour* and extract the plankton and whatever else is in the water along with it. From an egg, it takes a northern European mussel three to four years to slurp up enough plankton to reach marketable size; in the warmer, more plankton-rich waters of Spain, only about two years. Most edible oysters tend to grow a little slower, with water temperature and food supply key factors. In contrast, it is a rare chicken that goes to market when it is more than one year old.

But here's the thing: They are called farmed mussels and oysters, but the sea farmers are relying on wild plankton—free food—for their "crop." The farmers may sometimes pay a leasing fee, but they do not own the space occupied by their "farms," space that is typically public but with public access necessarily restricted. Access is also restricted for the other wildlife that would normally occupy the area filled with racks, pilings, cages, and lines. "Free food" is a relative term, too. What goes into the mollusks' tissues comes out of the ecosystem. Naturally occurring mollusks are part of that ecosystem, locking in and maintaining their piece of the carbon cycle, nitrogen cycle, and so on. Farmed mussels, clams, and oysters take those elements away when they are sold to distant markets.

Salmon farmers also have to provide food for their animals, and that usually means wild fish captured, ground, and formed into pellets. At an international aquaculture conference in Seattle, I asked one of the speakers what the salmon were fed, and he replied, "Well, we take fish that aren't doing anybody any good—anchovies from Peru and Chile, sometimes menhaden or any of those little fish that nobody eats, and we turn them into salmon." He said it takes four or five pounds of wild fish to make a pound of farmed salmon, but that doesn't account for the underlying investment in the fish that are fed to the salmon.

Although increasingly popular since the 1960s, salmon farming is controversial for a number of reasons. Crowding leads to diseases, and diseases lead to heavy use of antibiotics and chemical deterrents. Occasionally, fish escape, giving rise to additional concerns, such as the perils of interbreeding with and displacing native populations, as well as the possibility of spreading diseases. Other animals are affected as well: Birds, sea lions, seals, and dolphins are regarded as pests and "controlled" around salmon farms.

SUSTAINABLE AQUACULTURE

With all the worries and woes, there is little doubt that aquaculture is here to stay. New laws are being crafted and enacted that encourage use of state and federal waters in the United States and elsewhere for open-sea aquaculture ventures, and new technologies are being developed to support these initiatives. It is a race, of sorts, to see if in a few decades, new technologies, new policies, and new insights can be coupled with lessons learned over millennia to allow us to grow a few appropriate species as a source of sustenance. In effect, we need to extract food from the sea without actually taking more wildlife from the sea.

Critically important will be the task of developing closed systems away from the sea, systems in which everything, including the water, is recycled. At a conference in Australia on Fish, Aquaculture and Food Security in 2004, I made some proposals that are still relevant:

- Suppose, in the next decade, policies to govern both the Exclusive Economic Zones (marine zones exclusive to one country) and the High Seas (international waters) are implemented with an eye toward long-range sustainability rather than swift, short-term exploitation.
- Suppose we took a fraction of the annual cost of perverse subsidies for unsustainable fishing (currently more than $34 billion) and applied those resources to give fishermen other options.

- Suppose we applied another fraction and invested it in restoration of species and systems devastated by destructive fishing practices.
- Suppose we applied to aquaculture even one-tenth of the resources now being invested in agriculture.
- Suppose we seriously researched that great library, aquatic life, to identify a few species that could be adapted for cultivation, while understanding the vital roles that all animals have in maintaining a viable, healthy planet.
- Suppose we learned from 10,000 years of successes and failures in agriculture, and focused on developing a suite of aquatic organisms for cultivation that are:

  low on the food chain
  fast growing
  disease resistant
  tasty and nutritious or valuable for oils or other substances
  following suitable life cycles
  tolerant of crowding
  efficient converters of sunlight to plants or of plants to protein
- Suppose we focused more on closed-system cultivation of microbes, yeasts, and certain micro- and macro-algae as food sources for ourselves and the animals we grow—tiny organisms tailored to have desired features and high yields.
- Suppose we look at water issues, land and sea, recognizing that growing cows on an open pasture takes a huge amount of water, and so does raising salmon in open pens. It literally takes an ocean to raise a bluefin tuna.
- Suppose we accounted for the real water consumption involved in open or semi-closed systems, as compared to

recycling water in closed systems, and put that cost on the balance sheet.
- Suppose a priority for aquaculture—and agriculture—becomes more crop per drop.

Suppose these things don't happen.

Imagine the cost of doing business as usual.

Seaweed destined for the nest of this shag in the Falkland Islands.

PROTECTING THE OCEAN

11

Earth is a marine habitat.

—Nancy Foster, speaking at a meeting of the National Oceanic and Atmospheric Administration, 1990

I cannot remember ever being hotter or happier than on the night of August 30, 2002—the second time I had celebrated my birthday inside a submarine. Years before, I had occupied the observer's seat in the clear acrylic dome of one of the two *Johnson Sea-Link I* subs, watching its designer, Edwin A. Link, maneuver the manipulator arm on the second sub to pluck a few branches from a sponge and deliver them into my collecting tray as a Bahamian birthday bouquet.

Now I was alone, sitting inside the clear sphere of a *Deep Rover* submersible, feeling the temperature relentlessly rising above 38°C (100°F) within minutes of reaching the seafloor. The water temperature outside was a balmy 27°C (80°F), but five inches of acrylic retained heat generated by my body and the sub's instruments. That's fine when diving in deep water, where the outside temperature may be near freezing, even in the tropics, but my destination was little more than 30 meters (100 feet) down on a coral reef just offshore from the Mexican metropolis Veracruz in the Gulf of Mexico. I found an open, sandy space flanked by mounds and pillars of star coral, *Montastrea,* and settled in for 11 hours of dusk-to-dawn action.

It seemed unlikely that reefs could prosper in plain view of one of the country's busiest, most industrialized ports, home for more than 500,000 people, but John "Wes" Tunnell, a lean, lanky Texas ecologist who has dived the reefs of the western Gulf with classes of students for three decades, had chosen this place, Santiaguillo Reef, with care. Corals, sponges, and an amazing array of fish and invertebrate animals formed an underwater city of sorts, one where even the buildings were alive.

Designated years before as part of a network of protected areas along this part of Mexico's coast, my overnight residence was in better condition than most near-shore systems in the Gulf and Caribbean. Curtains of silver fish flowed over the reef, around the sub, then back to the reef, feasting on the clouds of plankton attracted to the sub's lights. A black-and-white-spotted moray slid by for a look, disappeared, then reappeared from time to time through the night. Notably missing were lobsters, groupers, and sharks—all critical components of healthy reef systems. Although protected, the Veracruz reefs have not yet fully recovered from years of heavy fishing pressure and other impacts that inhibit normal resilience.

On the surface, my daughter; Gale Mead, sub pilot and data manager for the project; scientists from the Harte Research Institute; and many of the cadets aboard the Mexican Navy's research vessel *Antares* stayed awake to hear my enthusiastic reports, communicated live from the reef below. Oceanographer and retired Mexican Navy admiral Alberto M. Vázquez de la Cerda, whose birthday coincides with mine, was eager to know if any of the corals were spawning. The night before, on a reef a few miles away, some of us had witnessed for the first time the synchronous release of golden eggs and clouds of sperm in the western Gulf. "No," I reported back. "No sign of eggs tonight." The magnificent burst of life that we had witnessed, vital to renew reefs damaged by storms and other impacts, by now was adrift in the plankton miles from the parent corals. A year would pass before there would be another such event.

The Veracruz expedition of 2002 marked an unusual collaboration of individuals, institutions, companies, and government agencies from two countries, focused on exploring, documenting, and ultimately protecting areas in the ocean that are still healthy, and restoring places that have been degraded. The project provided for me a model of how to overcome complexities involved with safeguarding even a small part of the ocean.

The Harte Research Institute at Texas A&M in Corpus Christi worked with NOAA, the National Geographic Society, and Mexican colleagues, notably Admiral Vázquez, to weave through the intricacies required to work in Mexican waters. We were encouraged to persevere by the institute's principal benefactor, Edward H. Harte, a pioneering Texas businessman and conservationist, who with his brother, Houston Harte, donated their 66,000-acre ranch in the foothills of the Rosillos Mountains to the Big Bend National Park, and who also was instrumental in the establishment of Mustang Island State Park and Padre Island National Seashore, among other significant actions aimed at protecting natural areas in Texas and beyond.

Establishing parks and protecting them on the land is challenging enough, but doing something similar in the ocean requires liberal doses of added ingenuity, and plenty of patience.

TWO GREAT IDEAS

The year 1872 was momentous in terms of the relationship of humankind to both land and sea. On land, the spectacular Yellowstone country in the Wyoming and Montana Territories was designated by the U.S. Congress as a "public park or pleasuring ground for the benefit and enjoyment of the people." It was the world's first national park. Others followed. Some—Sequoia, Yosemite, Mount Rainier, Glacier and Crater Lakes—were protected for their natural features. Others, including the dramatic cliff dwellings of Colorado's Mesa Verde area, were protected primarily to preserve the nation's historic and cultural heritage. By 1916, 14 national parks and 21 national monuments formed the basis for the system of national parks that some people call "the best idea America ever had."

For the ocean, 1872 marked the departure of the H.M.S. *Challenger* from England on the historic four-year global expedition of exploration that established the science of oceanography and forever changed attitudes about the nature of the sea. Exactly 100 years later, the U.S. Congress passed legislation authorizing the first underwater equivalent of national parks, national marine sanctuaries, to provide a measure of security for natural, historic, and cultural values under the sea—another brilliant idea.

While space must be made for people to live, farm, build cities, and otherwise do what people do to prosper, children today and the children of tomorrow can walk through redwood forests little changed in hundreds of years except for the fact that that the trees are taller and have greater girth. They can peer over the rim of the Grand Canyon and *not* see neon lights below, and can view the geysers and boiling springs of Yellowstone in the presence of buffalo. Nearly 400 places of natural, historic, and cultural significance are now within the U.S. National Park Service, and globally, thousands of other areas making up about 13 percent of the land have been designated for care. Fourteen national marine sanctuaries and a number of national marine monuments now embrace more than 881,000 square kilometers (340,000 square miles) of ocean in U.S. waters, and globally, nearly 5,000 marine protected areas have been

Like giant butterflies, golden rays gather in the Galápagos Islands.

designated. It sounds like a lot, but the total area of ocean covered is a fraction of 1 percent. That means, of course, that more than 99 percent is open for uses that are not entirely benign.

Some parks, sanctuaries, and monuments are remote, accessible to a hardy few, and generally take care of themselves, whereas others, especially those close to urban areas, must be safeguarded carefully to make it possible to accommodate the affections of multitudes of admirers. Few people have been to all of them, and many have not physically set foot or flipper on or in any of them, but everyone, everywhere, derives vital benefits from their existence as reservoirs of biodiversity, havens for depleted and endangered species, a source of critically important resilience in the face of swift changes in climate, weather, and planetary chemistry.

The United States, like most other countries with a coastline, has jurisdiction of an Exclusive Economic Zone extending 322 kilometers (200 miles) seaward from the coast. Under the sea, there is another United States bounded by water and larger than the familiar region above the high-tide line. U.S. waters hold a mostly unexplored and unprotected treasury of peaks and valleys, canyons and plains, from coastal coral reefs, sea grass meadows, and kelp forests to the unique high-pressure environment of the ocean's greatest depth.

In the 1970s, Australia's Great Barrier Reef Marine Park Authority began by zoning broad areas of the 2,414-kilometer-long (1,500-mile) reef system for various uses, including about 3 percent of the area where fishing was restricted. In the U.S. the first national marine sanctuary was a small area designated in 1975 around the remains of the sunken Union warship *Monitor,* resting in more than 60 meters (200 feet) of water off the coast of North Carolina. In the same year, a small area of coral reefs at Key Largo became a national sanctuary, expanded in 1990 to encompass most of the 7,250 square kilometers (2,800 square miles) around the Florida Keys. The thousands of marine protected areas that now occur are mostly in coastal waters. Some are huge, such as the 363,000-square-kilometer (140,000-square-mile) Northwest Hawaiian National Marine Sanctuary and 388,000 square kilometers (150,000 square miles) of ocean designated by the island nation Kiribati, but most are small, some embracing less than a square kilometer. Altogether, they make up just 0.8 of 1 percent of the ocean.

SUSTAINABLE SEAS EXPEDITIONS

In a momentous telephone call in 1997, Duane Silverstein, then executive director of the Richard and Rhoda Goldman Fund, outlined the fund's interest in supporting a project that could encompass exploration, research, conservation, and education. Would I be interested in suggesting a project? Would I ever!

I had recently accepted an appointment as Explorer-in-Residence at the National Geographic Society, and with its full support, a cooperative agreement was forged with the Society, the Goldman Fund, and NOAA that set in motion a five-year program—the Sustainable Seas Expeditions—focusing on the nation's system of marine sanctuaries. Francesca Cava, recently retired captain in the NOAA Corps and formerly head of the National Marine Sanctuary Program, agreed to help organize and wrangle the operations. Coincidentally, my longtime friend Phil Nuytten was putting the final touches on a new "personal submersible," *Deep Worker,* a petite underwater vehicle capable of taking almost anyone almost anywhere in depths as great as 610 meters (2,000 feet). It seemed the perfect vehicle to use to explore and document previously inaccessible places within the sanctuaries, and to scout for new sites that might be added.

More than a hundred institutions joined in the project, from universities that provided scientific staff and equipment to corporations—Oceaneering International, Kerr-McGee, Oryx Energy, American Marine Corporation, and Deep Ocean Exploration and Research (DOER) among them. Leading NOAA's participation was Nancy Foster, then head of the National Ocean Service, the division that included the marine sanctuary program and the ships that would be assigned to the project. It took thousands of elements meshing in some kind of chaotic harmony to make the program work, and Foster's calm leadership melted one roadblock after another in the critical first year of organizing staff, training scientists to be sub pilots, getting NOAA to agree for the first time ever to deploy submarines from their ships, and solving critical diplomatic issues inherent in the unique public-nonprofit-industry partnerships that evolved. In a tragic twist, the ocean lost one of its best friends and champions when Dr. Foster succumbed to cancer in 2000, just as the Sustainable Seas Expeditions (SSE) was hitting its stride.

Nearly 47,000 square kilometers (18,000 square miles) were then in U.S. coastal waters, from the 13,000-square-kilometer (5,000-square-mile) Monterey Bay National Marine Sanctuary to tiny patches of coral reef in the northern Gulf of Mexico, the Flower Garden Banks, and the even smaller reef area at Fagatele Bay, American Samoa. We began operations in Monterey, California, after weeks of training sessions with two of the little *Deep Worker* subs at a test tank provided by the Monterey Bay Aquarium Research Institute (MBARI). Within minutes of taking the controls, MBARI's director, Marcia McNutt, made it look easy to maneuver tight turns in the tank, a skill that most pilots eventually could master in the calm confines of the pool.

Offshore, it was a little trickier, but over five years, more than 100 individuals piloted a changing assortment of the little one-person *Deep Worker* subs and their larger one-person predecessor, *Deep Rover.* The subs were sometimes launched from a pier or in the large indoor pool at the Texas A&M facility at College Station. At sea, operations were conducted from four NOAA ships, two U.S. Navy vessels, the Mexican research vessel *Antares,* two private company ships, a research vessel from the state of Florida, another from the U.S. Geological Survey, and one more from the Mote Marine Laboratory. Numerous additional seacraft were involved, from dozens of inflatable rubber boats to splendid private yachts and dive boats that helped move people around.

Updates from the ocean by NOAA observers provided an online running account of SSE progress at sea, and on land a series of "Student Summits" enabled teachers and students opportunities at various ports of call to question SSE scientists, crawl in and out of the subs, and view some of the thousands of hours of high-quality videos collected. The videos, thousands of still photographs, and hundreds of pages of notes and observations poured into the Marine Sanctuary databases, providing valuable baseline information that will help monitor and assess change over time.

HOW MUCH PROTECTION IS ENOUGH?

National parks did not come into existence, nor are they managed, to maximize the number of wild birds, squirrels, bears, and big cats that can be killed for market. Curiously, though, establishing protected areas in the ocean is often justified on the basis that the abundance and diversity of marine life will increase inside and in time will "spill over" into adjacent areas, thus enhancing fishing opportunities. It does make sense as a management technique if fish and other ocean wildlife are valued primarily as commodities to be extracted, but not if their importance as vital "life-support" components is understood, and if the larger issues of ocean health and resilience are priorities.

The organization Ducks Unlimited has for many years successfully promoted protection for marshes and other critical habitats for ducks, geese, and other "game birds," primarily because of the desire to hunt them for sport, and occasionally for dinner. They understand that if wild ducks, geese, and other targeted species are to prosper in sufficient numbers for the purposes of hunting, it isn't realistic to take them anywhere at any time in any number by any and all possible means.

But taking a small number of wild creatures out of large populations while protecting—not destroying—the habitat is very different from the industrial-scale extraction of wildlife now taking place in the sea. These issues seemed to intrigue President George W. Bush during a conversation over dinner at the White House in April 2006.

Ocean explorer Jean-Michel Cousteau had just shown his film *Voyage to Kure* to about 50 guests who had been invited to hear about plans to establish a new marine sanctuary in the Northwestern Hawaiian Islands, the topic of Cousteau's presentation. First

Lady Laura Bush had visited the area and was deeply concerned about the enormous amount of trash that had accumulated at Midway and other remote islands. For more than ten years, various individuals and organizations—the Marine Conservation Biology Institute, Ocean Conservancy, the Pew Foundation, and numerous others—had gathered data and made a strong case for enhanced protection for the line of reefs and islands that arc more than a thousand miles seaward from the well-known and highly populated main islands of Hawaii.

Following the film, guests were invited to a buffet dinner with open seating around tables for six. I was a little slow getting through the line, and my friend and guest for the evening, Linda Glover, was already seated by the time I entered the room. President Bush had chosen to sit next to her, and next to him was Cousteau. The president of the National Marine Sanctuary Foundation, Lori Arguelles, and her husband moved to join them. With considerable trepidation, I took the last place at the table.

For the next hour and a half, we talked about ocean issues, about energy use, climate change, plastic debris in the sea, fishing practices—and the need to protect the Northwestern Hawaiian Islands. "For there to be fishermen, there must be fish," I said at one point. "For there to be fish, there must be places where they are safe. On land, marshes are protected to give ducks and geese havens where they can nest and raise their young. Flyways are respected, and there are strict limits on when and how many birds can be taken. In the sea, industrial fishing has reduced many species by more than 90 percent. The future for them, and for the ocean as a whole, is bleak unless there are safe places for wildlife in the sea, just as there are on the land."

President Bush was not aware that U.S. marine sanctuaries allow commercial and sport fishing, except in very small areas. "Why are they called sanctuaries, then?" he asked, and Lori

Arguelles explained that the designated areas are managed to serve multiple uses, from recreation to various commercial activities.

All of us urged the need to have larger "fully protected" areas in the sea, where "even the fish could be safe." As he left following dinner, the President called across the room to James Connaughton, chairman of the Council on Environmental Quality, "Jim, make it happen. I want to have full protection for the Northwestern Hawaiian Islands. No fishing."

Six weeks later, with Cousteau, the governor of Hawaii, and various dignitaries, I was invited to stand next to the President when he signed a document establishing not a new "sanctuary," but rather, the Papahanaumokuakea Marine National *Monument,* 362,000 square kilometers (140,000 square miles) of ocean where even the fish can swim in peace—and make more fish.

The new monument significantly increased the area of the ocean that has some form of protection, but the amount remained well under one percent of the sea.

Just how much of the ocean *should* be protected to maintain vital life-support functions, restore and hold steady populations of seriously depleted fish and other ocean wildlife, and cope with growing dead zones, ocean acidification, climate change, and massive amounts of pollution?

In 1980, the International Union for Conservation of Nature (IUCN) and the World Wildlife Fund collaborated in the development of a World Conservation Strategy. Although no specific areas or numbers were proposed at that time, marine protected areas were expected to be large enough to accomplish the following objectives:

- Maintain essential ecological processes and life-support system functions.
- Preserve genetic diversity.
- Ensure the sustainable utilization of species and ecosystems.

Twenty years later, during the United Nations Millennium Summit, goals for 2010 were signed by 147 heads of state from 189 countries endorsing action plans including 10 percent of the ocean by 2010. At the 2002 United Nations Conference on Environment and Development (UNCED) in Johannesburg, the number suggested was 12 percent by 2012.

Callum Roberts, an astute conservation biologist at the University of York, through scholarly analysis of the minimum it would take to restore and stabilize severely depleted areas of the sea, suggested 30 percent of the ocean must be off limits to fishing.

While not specifying how much, it was encouraging that at the World Oceans Conference in May 2009, more than 400 leading scientists from more than 76 countries signed the Manado Ocean Declaration including the following resolution:

> We resolve to further establish and effectively manage marine protected areas, including representative resilient networks, in accordance with international law, as reflected in UNCLOS [United Nations Convention on the Law of the Sea], and on the basis of available science, recognizing the importance of their contributions to ecosystem goods and services, and to contribute to the effort to conserve biodiversity, sustainable livelihoods and to adapt to climate change.

Since 2006, the United States has added 865,757 square kilometers (335,744 square miles) of ocean in the Pacific as national monuments, places where even the fish, lobsters, and shrimp are secure. Other nations have taken bold actions: enhanced protection for Australia's Great Barrier Reef, (now increased from 6 percent to 33 percent), certain New Zealand fjords, and more than 388,000 square kilometers (150,000 square miles) of pristine waters of the island nation Kiribati. In May 2009, South

Africa announced the addition of an impressive new marine protected area around the Southern Ocean's Prince Edward Islands about the size of Oklahoma. In May, a large part of the Savu Sea was dedicated for protection by Indonesia.

Nonetheless, the total area of the sea that is protected in 2009 remains less than one percent.

THE SCIENCE OF MARINE RESERVES

High on my short list of heroes for science-based ocean conservation is Graeme Kelleher, for 25 years the chairman of Australia's Great Barrier Reef Marine Park Authority, and for his entire career a proponent of marine protected areas. He has helped guide a global initiative to consider what is needed to maintain the health of the ocean through a system of protected areas, and developed working groups in all of the world's marine regions to identify priorities for the establishment and improved management of marine protected areas with particular emphasis on the protection of biodiversity. He spearheaded publication of a critical assessment of the issues in a document, *A Global Representative System of Marine Protected Areas,* listing the world's MPAs, and forming the foundation of the IUCN's marine program, now headed by a visionary biologist, Carl Gustaf Lundin, who continues to foster international cooperation for ocean science and conservation.

One of the divisions of the IUCN is the World Commission on Protected Areas, and within that is a Global Marine Programme headed by a brilliant British biologist, Dan Laffoley, who is determined to see the development of a global representative system of "effectively managed and lasting networks of marine protected areas." Through a carefully crafted plan of action, he is determined to "improve the coverage of what we have, obtain greater effectiveness of what we do, and sustain what we have into the future."

Penguins defy gravity in the southern ocean.

Over the years, government agencies, numerous individuals, and various conservation organizations have wrestled with the scientific rationale underlying ocean protection as the need to identify and protect areas of critical significance becomes increasingly urgent. The terminology used is often confusing, with "protected areas" and "sanctuaries" often providing little sanctity for the wildlife within. Recent research is demonstrating that "every fish counts" in maintaining the integrity of an area. Even recreational fishing, promoted by some as a relatively benign use of protected areas, can alter the nature of the system.

The international version of *The Science of Marine Reserves,* published by PISCO, the Partnership for Interdisciplinary Studies of Coastal Oceans, defines marine reserves as "ocean areas that are fully protected from activities that remove animals and plants or alter habitats, except as needed for scientific monitoring." Prohibited activities include fishing, aquaculture, dredging, and mining, while nondestructive swimming, diving, and boating are allowed. In studies of more than 124 marine reserves in temperate and tropical areas, increases were documented in biomass, density—the number in a given area—body size, and species diversity. Even small fully protected areas can make a measurable difference, but the benefits of larger reserves include coverage of more habitat types, greater diversity, and greater insurance against catastrophes, from storms to oil spills.

Marine biologist Enric Sala recently left a coveted tenure position at Scripps Institution of Oceanography to develop a major ocean initiative with National Geographic to pursue his dream of securing protection for pristine places in the ocean—while there is still time. His vision is shared by Greg Stone, head of the New England Aquarium's Global Marine Program, who is leading an effort to protect large "seascapes" globally. The World Wildlife Fund, the Nature Conservancy, and other organizations that have traditionally emphasized terrestrial conservation are increasingly "thinking blue."

Currently, the great majority of marine protected areas, however they are defined, are concentrated in coastal waters, designated by the state or country with jurisdiction. The open sea beyond, the "high seas," belongs to no one—and everyone. It remained largely intact through all preceding history, protected by its inaccessibility until technologies developed during World War II and later opened the ocean to anyone with the resources to get there, from industrial fishers and miners to those seeking sunken ships laden with treasure.

Kristina Gjerde, high seas policy advisor to the IUCN, is leading a global team to assess and protect critical areas beyond national jurisdiction. Having a few areas delineated can help stabilize ocean ecosystems, but overarching policies are needed to complement designation of specific places. High on the list of "most destructive" and "least justifiable" things that cause harm is using draggers or bottom trawls to capture ocean wildlife. For the past decade, the Deep Sea Conservation Coalition has represented scientists from 69 countries who have been supporting an initiative to have a moratorium on bottom trawling in international waters, to provide time to at least explore and evaluate areas before exploitation.

REASONS FOR HOPE

Knowing is the key to caring, and with caring there is hope that people will be motivated to take positive actions. They might not care even if they know, but they can't care if they are unaware. In March 2006, at a conference in Spain, I had an opportunity to salute Google Earth, a powerful new way of conveying knowledge about the world through real images of the entire planet via computer. During my 15-minute presentation about the ocean, I said how much I love using Google Earth. "I start high in the sky, then zoom down to explore my backyard, my neighbor's backyard, find coffee shops and national parks, even fly through the Grand Canyon," I said.

John Hanke, who heads Google Earth, sat in the front row, waiting his turn to speak. I thanked him for making possible this amazing new way of seeing the world, and without really thinking about it, added, "But when are you going to finish it? You've done a great job with the land—*Google Dirt.* What about the ocean?" Rather than taking offense, after our presentations, Hanke invited me to visit the Googleplex in Mountain View, California, to explain what was missing. With a big smile I said, "Just most of the planet."

Most obvious was the need to show the ocean's mountains, valleys, broad plains, and deep trenches by incorporating a visualization of the seafloor—the bathymetry. The U.S. Navy maintains a massive global database at its facility in Mississippi, and visits there were arranged by Linda Glover, who as a civilian geologist has worked with naval oceanographers for several decades. A three-year collaboration with the Navy, DOER Marine, and Google followed, with advice provided by 30 experts enlisted from around the world to help assemble appropriate content. At the same time, Linda and I began the monumental job of pulling together data and writing text for the National Geographic *Ocean*: *An Illustrated Atlas,* including new maps of the seafloor that incorporate the best information that is publicly available. As new data are acquired, they will be incorporated electronically into Google Earth, and eventually, in online editions of the maps from the atlas *Ocean.*

"My dream is to be able to allow people to dive into the virtual ocean," I told Hanke. "Imagine being able to touch a little blue dot on the surface of the sea in Google Earth that will open the way to see what's underwater, to see what whales see, to have images of whales and the other creatures that most people will never get to see for themselves."

Layers of data were assembled, dozens of "Googlers" became involved, with Steven Miller and Jenifer Austin Foulkes consistently moving things along. Finally, on February 2, 2009, at the California Academy of Science in San Francisco, Vice President Al Gore, Google CEO Eric Schmidt, John Hanke, and I spoke, Jimmy Buffett sang, we all smiled, and Google Earth was made whole.

Later that week, at the Technology, Entertainment, and Design (TED) Conference in Long Beach, California, I was given a chance to articulate "one wish that could change the world," and the Tedsters would mobilize to help make it come true. Since the previous October, when TED's "curator," Chris Anderson, called with the news that I had been chosen for a TED Prize, I had fretted nonstop, but not about what the wish would be. That was the easy part. I was concerned about how to convey convincingly in 18 minutes flat the reasons why it matters that the world is blue, why having networks of protected areas in the sea is vital—not just for the fish and whales—but for us, for the survival of all we hold near and dear, including life itself.

I call my wish list of protected areas in the sea "hope spots," and the list is long. Some are large, such as all of the Arctic Ocean, never touched throughout human history, now vulnerable to exploitation; the Coral Sea, the largely pristine ocean bordering the Great Barrier Reef; the large "Coral Triangle" of reefs and deep water near Indonesia; the Sargasso Sea, a vast region of open water near Bermuda that harbors a floating forest of golden brown *Sargassum* seaweed, crowded with unique forms of life; the southern ocean, especially the Ross Sea, still largely intact, despite increasing inroads by commercial fishing; the Galápagos Islands, designated as World Heritage areas, land and sea, but currently declining owing to fishing, both legal and illegal. Some of the hope spots are small, gems such as the "topographic highs" in the northern Gulf of Mexico and Pulley Ridge, an ancient barrier island submerged 240 kilometers (150 miles) west of Naples, Florida, or Drakes Estero, the most pristine bay in California, now being claimed for ongoing oyster farming. Not all are healthy, including Chesapeake Bay, where protection for the menhaden, crabs, oysters, and others wildlife could bring about profound improvement.

At TED, I said that a Global Plan of Action with the IUCN is under way to protect biodiversity, to mitigate and recover from the impacts of climate change; on the high seas and in coastal areas, wherever we can identify critical places. New technologies are needed to map, photograph, and explore the 95 percent of the ocean that we have yet to see.

And this is my wish:

> I wish you would use all means at your disposal—Films! Expeditions! The Web! New submarines!—to create a campaign

to ignite public support for a global network of Marine Protected Areas, "hope spots" large enough to save and restore the ocean, the blue heart of the planet.

How much? Some say 10 percent, some say 30 percent. You decide: How much of your heart do you want to protect? Whatever it is, a fraction of one percent is not enough.

In effect, the TED wish is the shorthand version of all that is in this book:

Throughout the history of our species, the mostly blue planet has kept us alive. It's time for us to return the favor.

ACKNOWLEDGMENTS

Deepest thanks to all who have contributed to this book, and even deeper apologies for errors and omissions that may have crept into what is recorded here, despite my best efforts and the admirable skills of the National Geographic's editorial team. Special thanks to Barbara Brownell Grogan, who, more than any other, made this book possible; to Nina Hoffman, who provided her consistent encouragement; and to Terry Garcia and John Fahey, who have never flinched in their support, no matter what (so far).

The World Is Blue represents the distillation of many decades of absorbing what I could of experiences that have shaped what is expressed here, starting with the ethic conveyed by my parents, Lewis R. Earle and Alice Richie Earle, and my brothers, Lewis S. Earle and Evan C. Earle, who shared early years and continue to guide my thinking. To my family—Elizabeth, Ian, Taylor, Morgan, Richie, Tamara, Russell, Kevin, and Gale—I credit the inspiration to write and the amazing experiences shared, above and below the ocean; and I send thanks for putting up with the benign neglect that writing requires, time that I long to have spent with you.

Thanks, too, to my extended family, those who have been especially close to me during years of exploring the ocean and sharing dreams about how to make a difference—Joan Membery, Linda Glover, Elaine Harrison, Kathy Sullivan, Shari Sant Plummer, Ellen Prager, Robert Wicklund; and to individuals who may or may not realize how profoundly they got under my skin with their insistence on the importance of exploring the natural world and conveying the truth in what is discovered—Harold J. Humm, Catherine Bowen, Edna Turnure, Jacques Cousteau, Edwin A.

Link, Sir Peter Scott, Sir David Attenborough, Sir John Rawlins, Roger Payne, Carl Safina, Jane Lubchenco, Sven Lindblad, Tim Kelly, Chris Parsons, Francesca Cava, John Robinson, Bruce Robison, Edith Widder, John Craven, Rita Colwell, Wes Tunnell, Julie Packard, Marcia McNutt, Jack Dangermond, John Hanke, Jake Ebert, Loran Fraser, the artist known as Wyland, and, although I never knew them personally, William Beebe, Edward Ellsberg, Thomas Huxley, Aldo Leopold, Loren Eisley, and hundreds of others who, through their writings, made me realize the importance of sharing experiences with those who will be around beyond my own time. And thanks to Ed Harte and his wonderful family, and to Robert Furgason, Wes Tunnell, and the entire Harte Research Institute team for their ethic of caring and years of collaboration. And last, thanks to my colleagues who share the dream embodied in the Deep Search Foundation; to those at Google, especially John Hanke, Eric and Wendy Schmidt, Michael T. Jones, Steve Miller, Jenifer Austin Foulkes, Rebecca Moore, and founders Larry Page and Serge Brin; and to the team at TED, especially Chris Anderson, and all of the Tedsters for backing the vision—and the reality—of *The World Is Blue.*

ABOUT THE AUTHOR

Called Her Deepness by the *New Yorker* and the *New York Times* and a Living Legend by the Library of Congress, and named by *Time* magazine as the first Hero for the Planet, Sylvia Earle is an oceanographer, explorer, author, lecturer, Explorer-in-Residence at the National Geographic Society, leader of the Sustainable Seas Expeditions, council chair for the Harte Research Institute for Gulf of Mexico Studies at Texas A&M-Corpus Christi, founder and chairman of the Deep Search Foundation, and chair of the Advisory Council for the Ocean in Google Earth. Formerly the chief scientist of the National Oceanic and Atmospheric Administration, she has founded three companies, including Deep Ocean Exploration and Research (DOER Marine Operations), and has served on various corporate boards, including the boards of Dresser Industries, Kerr-McGee, Oryx Energy Company, and Undersea Industries. She is a graduate of St. Petersburg College and Florida State University, with an M.A. and a Ph.D. from Duke University, and has received 17 honorary doctorates.

Author of more than 175 publications, lecturer in more than 70 countries, and participant in numerous television and radio productions, Earle has made the focus of her research the ecology and conservation of marine ecosystems, with special reference to marine algae and development of technology for access to and research in the deep sea. Worldwide field experience includes leading more than 100 expeditions and logging more than 7,000 hours underwater, including nine saturation dives and use of various submarines. She serves on the boards of organizations including the Woods Hole Oceanographic Institution, the Mote Marine

Laboratory, the Rutgers Institute of Marine and Coastal Sciences, the Conservation Fund, the Aspen Institute, the Ocean Futures Society, American Rivers, the Ocean Conservancy, and the Marine Conservation Biology Institute. She is a patron of Wildscreen, co-chairs the Science Committee of the U.S. National Parks Second Century Commission, and is a member of the Aspen Institute's Arctic Climate Change Commission.

She has received more than 100 national and international honors, including the National Women's Hall of Fame, the Netherlands' Order of the Golden Ark, Australia's Banksia Award, Italy's Premio Artiglio Award, and medals from the Explorers Club and the Society of Women Geographers. In 2009 she received the TED Prize, the Audubon Society's Rachel Carson Award, the BLUE Ocean Film Lifetime Achievement Award, and was inducted into the International Women's Forum.

TED TALK

One Wish to Change the World. A speech given at the 2009 Technology, Entertainment, and Design (TED) Conference

Fifty years ago, when I began exploring the ocean, no one, not Jacques Perrin, nor Jacques Cousteau, nor Rachel Carson, imagined that we could harm or do anything to the ocean by what we put in or took out of it. It seemed to be a Sea of Eden. But now we know we are facing "Paradise Lost."

I want to share with you my personal view of changes in the sea that affect all of us: to consider why it matters that in 50 years we have taken—eaten—more than 90 percent of the big fish in the sea; why you should care that nearly half of the coral reefs have disappeared; why a mysterious depletion of oxygen in large areas of the Pacific should concern not only the creatures that are dying, but it should concern you—why it does concern you as well.

I'm haunted by what [businessman and conservationist] Ray Anderson calls Tomorrow's Child asking why we didn't do something on our watch to save sharks, bluefin tuna, squids, and coral reefs—the living ocean—while there was still time. Well, now *is* that time.

I hope for your help to explore and protect the wild ocean in ways that will restore the health of the sea—and in so doing, secure hope for humankind. The health of the ocean means health for us. And I hope [SETI director] Jill Tarter's wish to engage Earthlings includes dolphins, whales, and other sea creatures in this quest to look for intelligent life elsewhere in the universe. I hope, Jill, that someday you will find evidence that there is intelligent life among humans on this planet. (Did I say that? I guess I did.)

For me, as a scientist, it all began in 1953 when I first tried scuba; it's when I first got to know fishes swimming in something other than lemon slices and butter. I actually love diving at night; you see a lot of fish then that you don't see in daytime. Diving day and night was really easy for me in 1970, when I led a team of aquanauts living underwater for weeks at a time, at the same time that astronauts were putting their footprints on the moon.

In 1979, I had a chance to put my footprints on the ocean floor while using the personal submersible called Jim—six miles offshore from Hawaii and 1,250 feet down. It's one of my favorite bathing suits. Since then, I've used more than 30 kinds of submarines, even started three companies and a nonprofit foundation called Deep Search, to design and build systems to access the deep sea. I led a five-year National Geographic program—the Sustainable Seas Expeditions—using little subs. They're so simple to drive that even scientists can do it, and I'm living proof.

Astronauts and aquanauts really appreciate the importance of air, food, water, and temperature—all the things you need to stay alive in space or under the sea. I heard astronaut Joe Allen, explain how he had to learn everything he could about his life-support system, and then do everything he could to take care of it. Then he pointed to this, our blue planet, and said, "Life-support system." We need to learn everything we can about it, and do everything we can, to take care of it. The poet W. H. Auden said, "Thousands have lived without love—not one without water."

Ninety-seven percent of Earth is ocean. No blue, no green. If you think the ocean isn't important, imagine Earth without it. Mars comes to mind. No ocean, no life-support system. I gave a talk not long ago at the World Bank and I showed this amazing image of Earth and said, "There it is: the World Bank. That's where all the assets are." And we have been drawing them down much faster than the natural systems can replenish them. [President of the UN Foundation] Tim Wirth says the economy is a wholly owned subsidiary of the environment. With every drop of water you drink, with every breath you take, you are connected to the sea, no matter where on Earth you live.

Most of the oxygen in the atmosphere is generated by life in the sea. Over time, most of the planet's organic carbon has been absorbed and stored there, mostly by microbes. The ocean drives climate and weather, stabilizes temperature, and shapes the Earth's chemistry. Water from the sea forms clouds that return to the land and sea as rain, sleet, and snow. And the ocean provides a home for about 97 percent of life in the world, maybe in the universe. No water? No life. No blue, no green. Yet we have this idea, we humans, that the Earth—all of it—the oceans, the skies, are so vast, so resilient, that it doesn't matter what we do to it. That may have been true 10,000 years ago, and maybe even 1,000 years ago. But in the last 100, especially the last 50, we have drawn down the assets—the air, the water, the wildlife—that make our lives possible.

New technologies are helping us to understand the nature of nature—the nature of what's happening, showing us our impact on the Earth. First, you have to know that you have a problem. And fortunately, in our time, we've learned more about the problems than in all preceding history. With knowing comes caring; with caring there's hope that we can find an enduring place for ourselves within the natural systems that support us. But first we have to know.

Three years ago, I met John Hanke, who heads up Google Earth, and told him how much I love being able to hold the world in my hands and go exploring vicariously. But, I asked, "When are you going to finish it? You've done a great job with the land—Google dirt—but what about the water?" I guess John took my question to heart, and since then I've had the great pleasure of working with the Googlers, DOER Marine [a marine engineering firm originally known as Deep Ocean Exploration and Research], the National Geographic, and dozens of the best institutions and scientists around the world—ones that we could enlist—to put the ocean in Google Earth. As of last Monday, February 2, Google Earth is now whole.

Consider this—starting right at the Long Beach Convention Center, we can find the aquarium, and then cruise up the coast to the big aquarium—the Pacific Ocean—and California's four national marine sanctuaries—and a new network of state marine reserves that are beginning to protect and restore some of the assets.

We can flit over to Hawaii—and see the real Hawaiian Islands. Not just the little bit that pokes through the ocean's surface, but also what's under the surface, to see what the whales see. We can explore the other side of the Hawaiian Islands. We can visit with the humpback whales, gentle giants that I've had the pleasure of meeting face-to-face many times underwater. There is nothing quite like being personally inspected by a whale.

We can pick up and fly to the ocean's deepest place, seven miles down, the Mariana Trench, where only two people have ever been. Imagine that—it's only seven miles but only two people have been there, and that was 49 years ago. One-way trips are easy. We need new deep-diving submarines. How about an X Prize for ocean exploration? We need to visit the deep trenches and the undersea mountains, to understand life in the deep sea.

Now we can go to the Arctic. Just ten years ago, I stood on ice at the North Pole. An ice-free Arctic Ocean may happen in this century. That's bad news for polar bears. That's bad news for us too. Excess carbon dioxide is not only driving global warming; it is also changing ocean chemistry, making the sea more acidic. That's bad news for coral reefs and oxygen-producing plankton—and also bad news for us.

We are putting hundreds of millions of tons of plastic and other trash into the sea, millions of tons of discarded fishing nets and gear that continue to kill. We are clogging the ocean, poisoning the planet's circulatory system. And we are taking out hundreds of millions of tons of wildlife—all carbon-based units. Barbarically, we're killing sharks for shark fin soup.

Food chains shape planetary chemistry, drive the carbon cycle, the nitrogen cycle, the oxygen cycle, and the water cycle—our life-support system. Incredibly, we're still killing bluefin tuna, truly endangered, and much more valuable alive than dead.

All of these are parts of our life-support system. We kill using longlines with baited hooks every few feet that may stretch for 50 miles or more. Industrial trawlers and draggers scrape the seafloor like bulldozers, taking everything in their path. Using Google Earth, you can witness trawlers in China, the North Sea, and the Gulf of Mexico, shaking the foundation of our life-support system, leaving plumes of death in their path.

The next time you dine on sushi or sashimi or a swordfish steak or shrimp cocktail, or whatever wildlife from the ocean you happen to enjoy, think of the real cost. For every pound that goes to market, more than 10 pounds—even 100 pounds—may be thrown away as bycatch. This is the consequence of not knowing that there are limits to what we can take out of the sea. This chart shows the decline in ocean wildlife from 1900 to 2000. The highest concentrations are in red. Imagine, in my lifetime, 90 percent of the big fish have been killed, and most of the turtles, sharks, tunas, and whales are way down in numbers. But there is good news: 10 percent of the big fish still remain. There are still some blue whales, there are still some krill in Antarctica, and a few oysters in Chesapeake Bay. Half of the coral reefs are still in pretty good shape—a jeweled belt around the middle of the planet.

There is still time, but not a lot, to turn things around. But business as usual means that in 50 years there may be no coral reefs and no commercial fishing, because the fish will simply be gone. Imagine the ocean without fish. Imagine what that means to our life-support system. Natural systems on the land are in big trouble, too, but the problems are more obvious—and some actions are being taken to protect trees, watersheds, and wildlife.

In 1872, with Yellowstone National Park, the United States began establishing a system of parks, which some say is "the best idea America ever had." About 12 percent of the land around the world is now protected—safeguarding biodiversity, providing a carbon sink, generating oxygen, and protecting watersheds.

In 1972, this nation began to establish a counterpart in the sea: national marine sanctuaries. It's another *great* idea. The good news is that there are now more than 4,000 places in the sea around the world that have some protection, and you can find them on Google Earth. The bad news is that you have to look hard to find them. In the last three years, for example, the United States protected 340,000 square miles of ocean, as national monuments. But that only increased from 0.6 percent to 0.8 percent of the ocean that's protected globally.

Protected areas do rebound, but it takes a long time to restore 50-year-old rockfish or monkfish, or sharks, or sea bass, or 200-year-old orange roughy. We don't consume 200-year-old cows or chickens. Protected areas provide hope—hope that the creatures of [biologist E. O.] Wilson's dream of an Encyclopedia of Life, a census of marine life—will live, not just as a list, a photograph, or a paragraph.

With scientists around the world, I've been looking at the 99 percent of the ocean that remains open to fishing, mining, drilling, dumping, and whatever—to search out "hope spots" and try to find ways to give them, and us, a secure future. Such as the Arctic. We have one chance right now to get it right. Or the Antarctic, where the continent is protected, but the surrounding ocean is being stripped of its krill, whales, and fish.

The Sargasso Sea's three million square miles of floating forest is being gathered up to feed cows. Ninety-seven percent of the land in the Galápagos Islands is protected, but the adjacent sea is being ravaged by fishing. It's true, too, in Argentina, where the Patagonian shelf is now in serious trouble. It's true in the high seas, where whales, tunas, and turtles travel; the largest, least protected ecosystem on Earth, filled with luminous creatures living in dark waters that average two miles deep. They flash and sparkle and glow with their own living light.

There are still places in the sea as pristine as I knew as a child. The next 10 years may be the most important in the next 10,000 years—the best chance our species will have to protect what remains of the natural systems that give us life. To cope with climate change we need new ways to generate power. We need better ways to cope with poverty, wars, and disease. We need many things to keep and maintain the world as a better place. But nothing else will matter if we fail to protect the ocean. Our fate, and the ocean's, are one. We need to do for the ocean what Al Gore did for the skies above.

A Global Plan of Action with the IUCN—the International Union for Conservation of Nature—is under way to protect biodiversity, to mitigate and recover from the impacts of climate change, on the high seas and in coastal areas, wherever we can identify critical places. New technologies are needed to map, photograph, and explore the 95 percent of the ocean that we have yet to see. Biodiversity provides stability and resilience. We need deep-diving subs, new technologies to explore the ocean. Maybe we need an expedition—a TED at sea—that could help figure out our next steps.

And, so, I suppose you want to know what my wish is:

I wish you would use all means at your disposal—Films! Expeditions! The Web! New submarines!—to create a campaign to ignite public support for a global network of Marine Protected Areas, hope spots large enough to save and restore the ocean, the blue heart of the planet.

How much? Some say 10 percent. Some say 30 percent. You decide: How much of your heart do you want to protect? Whatever it is, a fraction of one percent is not enough.

My wish is a big wish, but if we can make it happen, it truly can change the world and help ensure the survival of what is, as it turns out, actually my favorite species: human beings.

For the children of today, for Tomorrow's Child, as never again, *now* is the time.

PROTECTED MARINE SITES

From: Charter Members of the National System of Marine Protected Areas (April 2009) Marine Protected Areas of the United States. http://mpa.gov/national_system/nationalsystem_list.html

Site Name	State	Government	Management Agency
ACE Basin National Wildlife Refuge	South Carolina	Federal	U.S. Fish and Wildlife Service
Admiralty Head Marine Preserve	Washington	State	Washington Department of Fish and Wildlife
Ahihi-Kinau Natural Area Reserve	Hawaii	State	Hawaii Department of Land and Natural Resources
Alaska Maritime National Wildlife Refuge	Alaska	Federal	U.S. Fish and Wildlife Service
Alligator River National Wildlife Refuge	North Carolina	Federal	U.S. Fish and Wildlife Service
Anahuac National Wildlife Refuge	Texas	Federal	U.S. Fish and Wildlife Service
Ano Nuevo ASBS State Water Quality Protection Area	California	State	California State Water Resources Control Board
Ano Nuevo State Marine Conservation Area	California	State	California Department of Fish and Game
Aransas National Wildlife Refuge	Texas	Federal	U.S. Fish and Wildlife Service
Arctic National Wildlife Refuge	Alaska	Federal	U.S. Fish and Wildlife Service
Argyle Lagoon San Juan Islands Marine Preserve	Washington	State	Washington Department of Fish and Wildlife
Asilomar State Marine Reserve	California	State	California Department of Fish and Game
Assateague Island National Seashore	Virginia & Maryland	Federal	National Park Service
Aua	American Samoa	Territorial	Department of Marine and Wildlife Resources
Back Bay National Wildlife Refuge	Virginia	Federal	U.S. Fish and Wildlife Service
Baker Island National Wildlife Refuge	Pacific Islands	Federal	U.S. Fish and Wildlife Service
Bandon Marsh National Wildlife Refuge	Oregon	Federal	U.S. Fish and Wildlife Service
Bethel Beach Natural Area Preserve	Virginia	State	Virginia Department of Conservation and Recreation
Big Boggy National Wildlife Refuge	Texas	Federal	U.S. Fish and Wildlife Service
Big Branch Marsh National Wildlife Refuge	Louisiana	Federal	U.S. Fish and Wildlife Service
Big Creek State Marine Conservation Area	California	State	California Department of Fish and Game
Big Creek State Marine Reserve	California	State	California Department of Fish and Game
Bird Rock ASBS State Water Quality Protection Area	California	State	California State Water Resources Control Board
Biscayne National Park	Florida	Federal	National Park Service

Year Established	Level of Protection	Primary Conservation Focus	Site Area (km^2)
1990	Uniform Multiple Use	Sustainable Production	78.2
2002	Uniform Multiple Use	Natural Heritage	0.4
1973	No Impact	Natural Heritage	3.2
1980	Uniform Multiple Use	Natural Heritage	26376.6
1984	Uniform Multiple Use	Natural Heritage	547.4
1963	Uniform Multiple Use	Natural Heritage	95.7
1974	Uniform Multiple Use	Natural Heritage	55.0
2007	Uniform Multiple Use	Natural Heritage	28.8
1937	Uniform Multiple Use	Natural Heritage	26.6
1960	Uniform Multiple Use	Natural Heritage	413.9
1990	Uniform Multiple Use	Natural Heritage	0.1
2007	No Take	Natural Heritage	3.9
1965	Zoned Multiple Use	Natural Heritage	124.4
2003	Uniform Multiple Use	Cultural Heritage	0.2
1938	Zoned Multiple Use	Natural Heritage	64.8
1974	No Access	Natural Heritage	130.1
1983	Uniform Multiple Use	Natural Heritage	1.2
1991	Uniform Multiple Use	Natural Heritage	0.0
1983	Uniform Multiple Use	Natural Heritage	39.1
1994	Uniform Multiple Use	Natural Heritage	110.8
2007	Uniform Multiple Use	Natural Heritage	20.4
1994	No Take	Natural Heritage	37.6
1974	Uniform Multiple Use	Natural Heritage	0.4
1968	Zoned Multiple Use	Natural Heritage	713.6

ASBS: Areas of Special Biological Significance
Note: "Site Area" includes marine area only

Site Name	State	Government	Management Agency
Blackwater National Wildlife Refuge	Maryland	Federal	U.S. Fish and Wildlife Service
Blake Island Underwater Park	Washington	State	Washington State Parks & Recreation Commission
Block Island National Wildlife Refuge	Rhode Island	Federal	U.S. Fish and Wildlife Service
Blue Crab Sanctuary	Virginia	State	Virginia Marine Resource Commission
Bodega ASBS State Water Quality Protection Area	California	State	California State Water Resources Control Board
Bombay Hook National Wildlife Refuge	Delaware	Federal	U.S. Fish and Wildlife Service
Bon Secour National Wildlife Refuge	Alabama	Federal	U.S. Fish and Wildlife Service
Brackett's Landing Shoreline Sanctuary Conservation Area	Washington	Partnership	Washington Department of Fish and Wildlife
Brazoria National Wildlife Refuge	Texas	Federal	U.S. Fish and Wildlife Service
Breton National Wildlife Refuge	Louisiana	Federal	U.S. Fish and Wildlife Service
Cambria State Marine Conservation Area	California	State	California Department of Fish and Game
Cape May National Wildlife Refuge	Delaware	Federal	U.S. Fish and Wildlife Service
Cape Romain National Wildlife Refuge	South Carolina	Federal	U.S. Fish and Wildlife Service
Carmel Bay ASBS State Water Quality Protection Area	California	State	California State Water Resources Control Board
Carmel Bay State Marine Conservation Area	California	State	California Department of Fish and Game
Carmel Pinnacles State Marine Reserve	California	State	California Department of Fish and Game
Cedar Island National Wildlife Refuge	North Carolina	Federal	U.S. Fish and Wildlife Service
Cedar Keys National Wildlife Refuge	Florida	Federal	U.S. Fish and Wildlife Service
Channel Islands National Marine Sanctuary	California	Federal	National Marine Sanctuaries
Channel Islands National Park	California	Federal	National Park Service
Chassahowitzka National Wildlife Refuge	Florida	Federal	U.S. Fish and Wildlife Service
Cherry Point Aquatic Reserve	Washington	State	Washington Department of Fish and Wildlife
Chincoteague National Wildlife Refuge	Virginia & Maryland	Federal	U.S. Fish and Wildlife Service
Conscience Point National Wildlife Refuge	New York	Federal	U.S. Fish and Wildlife Service
Cordell Bank National Marine Sanctuary	California	Federal	National Marine Sanctuaries
Crocodile Lake National Wildlife Refuge	Florida	Federal	U.S. Fish and Wildlife Service
Cross Island National Wildlife Refuge	Maine	Federal	U.S. Fish and Wildlife Service
Crystal River National Wildlife Refuge	Florida	Federal	U.S. Fish and Wildlife Service
Currituck National Wildlife Refuge	North Carolina	Federal	U.S. Fish and Wildlife Service
Cypress Island Aquatic Reserve	Washington	State	Washington Department of Natural Resources

Year Established	Level of Protection	Primary Conservation Focus	Site Area (km²)
1933	Uniform Multiple Use	Natural Heritage	0.0
1970	Uniform Multiple Use	Natural Heritage	0.5
1973	Uniform Multiple Use	Natural Heritage	0.3
1994	Uniform Multiple Use	Sustainable Production	2447.5
1974	Uniform Multiple Use	Natural Heritage	0.6
1937	Uniform Multiple Use	Natural Heritage	85.7
1980	Uniform Multiple Use	Natural Heritage	28.4
1970	No Take	Natural Heritage	0.2
1966	Uniform Multiple Use	Natural Heritage	63.4
1904	Uniform Multiple Use	Natural Heritage	70.2
2007	Uniform Multiple Use	Natural Heritage	16.2
1989	Uniform Multiple Use	Natural Heritage	73.3
1930	Uniform Multiple Use	Natural Heritage	119.2
1975	Uniform Multiple Use	Natural Heritage	6.4
1976	Uniform Multiple Use	Natural Heritage	5.5
2007	No Take	Natural Heritage	1.4
1964	Uniform Multiple Use	Natural Heritage	68.0
1929	Uniform Multiple Use	Natural Heritage	3.4
1980	Zoned w/No Take Areas	Natural Heritage	3813.7
1938	Uniform Multiple Use	Natural Heritage	477.9
1943	Uniform Multiple Use	Natural Heritage	149.7
2000	Uniform Multiple Use	Natural Heritage	12.4
1943	Zoned Multiple Use	Natural Heritage	59.4
1971	No Impact	Natural Heritage	0.2
1989	Zoned Multiple Use	Natural Heritage	1371.7
1980	No Access	Natural Heritage	29.3
1980	No Take	Natural Heritage	6.2
1983	Uniform Multiple Use	Natural Heritage	34.1
1984	Uniform Multiple Use	Natural Heritage	1.2
2000	Uniform Multiple Use	Natural Heritage	23.9

Site Name	State	Government	Management Agency
Dameron Marsh Natural Area Preserve	Virginia	State	Virginia Department of Conservation and Recreation
Deception Pass Underwater Park	Washington	State	Washington State Parks & Recreation Commission
Del Mar Landing ASBS State Water Quality Protection Area	California	State	California State Water Resources Control Board
Delta National Wildlife Refuge	Louisiana	Federal	U.S. Fish and Wildlife Service
Don Edwards San Francisco Bay National Wildlife Refuge	California	Federal	U.S. Fish and Wildlife Service
Double Point ASBS State Water Quality Protection Area	California	State	California State Water Resources Control Board
Dry Tortugas National Park	Florida	Federal	National Park Service
Dungeness National Wildlife Refuge	Washington	Federal	U.S. Fish and Wildlife Service
Duxbury Reef ASBS State Water Quality Protection Area	California	State	California State Water Resources Control Board
Eastern Neck National Wildlife Refuge	Maryland	Federal	U.S. Fish and Wildlife Service
Eastern Shore of Virginia National Wildlife Refuge	Virginia	Federal	U.S. Fish and Wildlife Service
Edward F. Ricketts State Marine Conservation Area	California	State	California Department of Fish and Game
Edwin B. Forsythe National Wildlife Refuge	New Jersey	Federal	U.S. Fish and Wildlife Service
Elkhorn Slough State Marine Conservation Area	California	State	California Department of Fish and Game
Elkhorn Slough State Marine Reserve	California	State	California Department of Fish and Game
Everglades National Park	Florida	Federal	National Park Service
Fagatele Bay National Marine Sanctuary	American Samoa	Federal	National Marine Sanctuaries
False Bay San Juan Islands Marine Preserve	Washington	State	Washington Department of Fish and Wildlife
False Cape State Park	Virginia	State	Virginia Department of Conservation and Recreation
Farallon Islands ASBS State Water Quality Protection Area	California	State	California State Water Resources Control Board
Farnsworth Bank ASBS State Water Quality Protection Area	California	State	California State Water Resources Control Board
Featherstone National Wildlife Refuge	Virginia	Federal	U.S. Fish and Wildlife Service
Fidalgo Bay Aquatic Reserve	Washington	State	Washington Department of Natural Resources
Fisherman Island National Wildlife Refuge	Virginia	Federal	U.S. Fish and Wildlife Service
Florida Keys National Marine Sanctuary	Florida	Federal	National Marine Sanctuaries
Flower Garden Banks National Marine Sanctuary	Texas	Federal	National Marine Sanctuaries
Friday Harbor San Juan Islands Marine Preserve	Washington	State	Washington Department of Fish and Wildlife
Gerry E. Studds/Stellwagen Bank National Marine Sanctuary	Massachusetts	Federal	National Marine Sanctuaries
Gerstle Cove ASBS State Water Quality Protection Area	California	State	California State Water Resources Control Board
Glacier Bay National Park & Preserve	Alaska	Federal	National Park Service

Year Established	Level of Protection	Primary Conservation Focus	Site Area (km²)
1998	Uniform Multiple Use	Natural Heritage	0.1
1970	Uniform Multiple Use	Natural Heritage	0.4
1974	Uniform Multiple Use	Natural Heritage	0.2
1935	Uniform Multiple Use	Natural Heritage	206.1
1972	Uniform Multiple Use	Natural Heritage	34.7
1974	Uniform Multiple Use	Natural Heritage	0.4
1935	Zoned w/No Take Areas	Natural Heritage	279.7
1915	Uniform Multiple Use	Sustainable Production	3.8
1974	Uniform Multiple Use	Natural Heritage	3.5
1962	Uniform Multiple Use	Natural Heritage	8.6
1984	Uniform Multiple Use	Natural Heritage	5.7
2007	Uniform Multiple Use	Natural Heritage	0.6
1939	Zoned w/No Take Areas	Natural Heritage	276.8
2007	Uniform Multiple Use	Natural Heritage	0.2
1980	No Take	Natural Heritage	3.9
1934	Zoned w/No Take Areas	Natural Heritage	3521.5
1986	Zoned w/No Take Areas	Natural Heritage	0.7
1990	Uniform Multiple Use	Natural Heritage	1.2
1966	Uniform Multiple Use	Natural Heritage	15.7
1974	Uniform Multiple Use	Natural Heritage	46.2
1974	Uniform Multiple Use	Natural Heritage	0.2
1978	Uniform Multiple Use	Natural Heritage	1.3
2000	Uniform Multiple Use	Natural Heritage	2.8
1969	Uniform Multiple Use	Natural Heritage	6.8
1990	Zoned w/No Take Areas	Natural Heritage	9900.9
1992	Zoned Multiple Use	Natural Heritage	146.2
1990	Uniform Multiple Use	Natural Heritage	1.7
1992	Uniform Multiple Use	Natural Heritage	2189.8
1974	Uniform Multiple Use	Natural Heritage	0.0
1925	Uniform Multiple Use	Natural Heritage	2371.3

Site Name	State	Government	Management Agency
Grand Bay National Wildlife Refuge	Alabama & Mississippi	Federal	U.S. Fish and Wildlife Service
Grays Harbor National Wildlife Refuge	Washington	Federal	U.S. Fish and Wildlife Service
Gray's Reef National Marine Sanctuary	Georgia	Federal	National Marine Sanctuaries
Great Bay National Wildlife Refuge	New Hampshire	Federal	U.S. Fish and Wildlife Service
Great White Heron National Wildlife Refuge	Florida	Federal	U.S. Fish and Wildlife Service
Greyhound Rock State Marine Conservation Area	California	State	California Department of Fish and Game
Guam National Wildlife Refuge	Guam	Federal	U.S. Fish and Wildlife Service
Guana Tolomato Matanzas National Estuarine Research Reserve	Florida	Partnership	Florida Department of Environmental Protection
Gulf of the Farallones National Marine Sanctuary	California	Federal	National Marine Sanctuaries
Hanauma Bay Marine Life Conservation District	Hawaii	Partnership	Hawaii Department of Land and Natural Resources
Haro Strait Special Management Fishery Area	Washington	State	Washington Department of Fish and Wildlife
Hawaiian Islands Humpback Whale National Marine Sanctuary	Hawaii	Federal	National Marine Sanctuaries
Heisler Park ASBS State Water Quality Protection Area	California	State	California State Water Resources Control Board
Howland Island National Wildlife Refuge	Pacific Islands	Federal	U.S. Fish and Wildlife Service
Hughlett Point Natural Area Preserve	Virginia	State	Virginia Department of Conservation and Recreation
Huron National Wildlife Refuge	Michigan	Federal	U.S. Fish and Wildlife Service
Irvine Coast ASBS State Water Quality Protection Area	California	State	California State Water Resources Control Board
Island Bay National Wildlife Refuge	Florida	Federal	U.S. Fish and Wildlife Service
Isle Royale National Park	Minnesota & Michigan	Federal	National Park Service
J.N. Ding Darling National Wildlife Refuge	Florida	Federal	U.S. Fish and Wildlife Service
Jacques Cousteau National Estuarine Research Reserve	New Jersey	Partnership	Rutgers University, Institute of Marine and Coastal Sciences
James V. Fitzgerald ASBS State Water Quality Protection Area	California	State	California State Water Resources Control Board
Jarvis Island National Wildlife Refuge	Pacific Islands	Federal	U.S. Fish and Wildlife Service
John H. Chafee National Wildlife Refuge	Rhode Island	Federal	U.S. Fish and Wildlife Service
Johnston Island National Wildlife Refuge	Pacific Islands	Federal	U.S. Fish and Wildlife Service
Jughandle Cove ASBS State Water Quality Protection Area	California	State	California State Water Resources Control Board
Julia Pfeiffer Burns ASBS State Water Quality Protection Area	California	State	California State Water Resources Control Board
Kahoolawe Island Reserve	Hawaii	Partnership	Hawaii Department of Land and Natural Resources

Year Established	Level of Protection	Primary Conservation Focus	Site Area (km²)
1992	Uniform Multiple Use	Natural Heritage	71.6
1990	No Take	Natural Heritage	5.7
1981	Uniform Multiple Use	Natural Heritage	57.4
1992	Uniform Multiple Use	Natural Heritage	4.3
1938	Uniform Multiple Use	Natural Heritage	838.1
2007	Uniform Multiple Use	Natural Heritage	31.1
1993	Zoned w/No Take Areas	Natural Heritage	121.8
1999	Uniform Multiple Use	Natural Heritage	262.0
1981	Zoned Multiple Use	Natural Heritage	3327.0
1967	No Impact	Natural Heritage	0.4
1972	Uniform Multiple Use	Sustainable Production	52.6
1992	Uniform Multiple Use	Natural Heritage	3555.0
1974	Uniform Multiple Use	Natural Heritage	0.1
1974	No Access	Natural Heritage	139.9
1997	Uniform Multiple Use	Natural Heritage	0.0
1905	Uniform Multiple Use	Natural Heritage	0.6
1974	Uniform Multiple Use	Natural Heritage	3.8
1908	No Access	Natural Heritage	0.1
1931	Uniform Multiple Use	Natural Heritage	2223.1
1945	Zoned Multiple Use	Natural Heritage	32.9
1998	Uniform Multiple Use	Natural Heritage	450.5
1974	Uniform Multiple Use	Natural Heritage	2.1
1974	No Access	Natural Heritage	152.8
1989	Uniform Multiple Use	Natural Heritage	3.8
1926	No Access	Natural Heritage	278.5
1974	Uniform Multiple Use	Natural Heritage	0.8
1974	Uniform Multiple Use	Natural Heritage	7.1
1993	Zoned Multiple Use	Cultural Heritage	202.9

Site Name	State	Government	Management Agency
Kealakekua Bay Marine Life Conservation District	Hawaii	State	Hawaii Department of Land and Natural Resources
Key West National Wildlife Refuge	Florida	Federal	U.S. Fish and Wildlife Service
King Range ASBS State Water Quality Protection Area	California	State	California State Water Resources Control Board
Kingman Reef National Wildlife Refuge	Pacific Islands	Federal	U.S. Fish and Wildlife Service
Kiptopeke State Park	Virginia	State	Virginia Department of Conservation and Recreation
La Jolla ASBS State Water Quality Protection Area	California	State	California State Water Resources Control Board
Laguna Point to Latigo Point ASBS State Water Quality Protection Area	California	State	California State Water Resources Control Board
Lewis and Clark National Wildlife Refuge	Washington & Oregon	Federal	U.S. Fish and Wildlife Service
Lovers Point State Marine Reserve	California	State	California Department of Fish and Game
Lower Suwannee National Wildlife Refuge	Florida	Federal	U.S. Fish and Wildlife Service
Mackay Island National Wildlife Refuge	Virginia & North Carolina	Federal	U.S. Fish and Wildlife Service
Marin Islands National Wildlife Refuge	California	Federal	U.S. Fish and Wildlife Service
Martin National Wildlife Refuge	Virginia	Federal	U.S. Fish and Wildlife Service
Mashpee National Wildlife Refuge	Massachusetts	Federal	U.S. Fish and Wildlife Service
Matlacha Pass National Wildlife Refuge	Florida	Federal	U.S. Fish and Wildlife Service
Maury Island Aquatic Reserve	Washington	State	Washington Department of Fish and Wildlife
Midway Atoll National Wildlife Refuge	Hawaii	Federal	U.S. Fish and Wildlife Service
Molokini Shoal Marine Life Conservation District	Hawaii	State	Hawaii Department of Land and Natural Resources
Monomoy National Wildlife Refuge	Massachusetts	Federal	U.S. Fish and Wildlife Service
Monterey Bay National Marine Sanctuary	California	Federal	National Marine Sanctuaries
Moro Cojo Slough State Marine Reserve	California	State	California Department of Fish and Game
Morro Bay State Marine Recreational Management Area	California	State	California Department of Fish and Game
Morro Bay State Marine Reserve	California	State	California Department of Fish and Game
National Key Deer Refuge	Florida	Federal	U.S. Fish and Wildlife Service
Natural Bridges State Marine Reserve	California	State	California Department of Fish and Game
Nestucca Bay National Wildlife Refuge	Oregon	Federal	U.S. Fish and Wildlife Service
Ninigret National Wildlife Refuge	Rhode Island	Federal	U.S. Fish and Wildlife Service
Nisqually National Wildlife Refuge	Washington	Federal	U.S. Fish and Wildlife Service

Year Established	Level of Protection	Primary Conservation Focus	Site Area (km²)
1969	Zoned w/No Take Areas	Natural Heritage	1.2
1908	Uniform Multiple Use	Natural Heritage	863.8
1974	Uniform Multiple Use	Natural Heritage	101.5
2001	No Access	Natural Heritage	1968.0
1992	Uniform Multiple Use	Natural Heritage	2.0
1974	Uniform Multiple Use	Natural Heritage	1.8
1974	Uniform Multiple Use	Natural Heritage	48.0
1972	Uniform Multiple Use	Sustainable Production	105.5
2007	No Take	Natural Heritage	0.8
1979	Uniform Multiple Use	Natural Heritage	341.3
1960	Uniform Multiple Use	Natural Heritage	29.9
1992	No Access	Natural Heritage	1.9
1995	Uniform Multiple Use	Natural Heritage	16.9
1995	No Access	Natural Heritage	26.1
1908	No Access	Natural Heritage	2.3
2000	Uniform Multiple Use	Natural Heritage	22.4
1988	Uniform Multiple Use	Natural Heritage	2365.3
1977	Zoned w/No Take Areas	Natural Heritage	0.4
1944	Zoned Multiple Use	Natural Heritage	29.6
1992	Zoned Multiple Use	Natural Heritage	13813.3
2007	No Take	Natural Heritage	0.5
2007	Zoned Multiple Use	Natural Heritage	7.9
2007	No Take	Natural Heritage	0.8
1954	Uniform Multiple Use	Natural Heritage	561.0
2007	No Take	Natural Heritage	0.6
1991	No Access	Natural Heritage	1.9
1970	Uniform Multiple Use	Natural Heritage	1.8
1974	Zoned w/No Take Areas	Natural Heritage	7.3

Site Name	State	Government	Management Agency
NOAA's Monitor National Marine Sanctuary	North Carolina	Federal	National Marine Sanctuaries
Nomans Land Island National Wildlife Refuge	Massachusetts	Federal	U.S. Fish and Wildlife Service
Northwest Santa Catalina Island ASBS State Water Quality Protection Area	California	State	California State Water Resources Control Board
Occoquan Bay National Wildlife Refuge	Virginia	Federal	U.S. Fish and Wildlife Service
Olympic Coast National Marine Sanctuary	Washington	Federal	National Marine Sanctuaries
Orchard Rocks Conservation Area	Washington	State	Washington Department of Fish and Wildlife
Oyster Bay National Wildlife Refuge	New York	Federal	U.S. Fish and Wildlife Service
Pacific Grove ASBS State Water Quality Protection Area	California	State	California State Water Resources Control Board
Pacific Grove Marine Gardens State Marine Conservation Area	California	State	California Department of Fish and Game
Palmyra Atoll National Wildlife Refuge	Pacific Islands	Federal	U.S. Fish and Wildlife Service
Papahanaumokuakea Marine National Monument	Hawaii	Partnership	National Marine Sanctuaries
Parker River National Wildlife Refuge	Massachusetts	Federal	U.S. Fish and Wildlife Service
Pea Island National Wildlife Refuge	North Carolina	Federal	U.S. Fish and Wildlife Service
Pelican Island National Wildlife Refuge	Florida	Federal	U.S. Fish and Wildlife Service
Piedras Blancas State Marine Conservation Area	California	State	California Department of Fish and Game
Piedras Blancas State Marine Reserve	California	State	California Department of Fish and Game
Pine Island National Wildlife Refuge	Florida	Federal	U.S. Fish and Wildlife Service
Pinellas National Wildlife Refuge	Florida	Federal	U.S. Fish and Wildlife Service
Plum Tree Island National Wildlife Refuge	Virginia	Federal	U.S. Fish and Wildlife Service
Point Buchon State Marine Conservation Area	California	State	California Department of Fish and Game
Point Buchon State Marine Reserve	California	State	California Department of Fish and Game
Point Lobos ASBS State Water Quality Protection Area	California	State	California State Water Resources Control Board
Point Lobos State Marine Conservation Area	California	State	California Department of Fish and Game
Point Lobos State Marine Reserve	California	State	California Department of Fish and Game
Point Reyes Headlands ASBS State Water Quality Protection Area	California	State	California State Water Resources Control Board
Point Reyes National Seashore	California	Federal	National Park Service
Point Sur State Marine Conservation Area	California	State	California Department of Fish and Game
Point Sur State Marine Reserve	California	State	California Department of Fish and Game
Pond Island National Wildlife Refuge	Maine	Federal	U.S. Fish and Wildlife Service

Year Established	Level of Protection	Primary Conservation Focus	Site Area (km^2)
1975	Uniform Multiple Use	Cultural Heritage Restrictions	2.2
1970	No Access	Natural Heritage	2.5
1974	Uniform Multiple Use	Natural Heritage	53.7
1973	Uniform Multiple Use	Natural Heritage	0.3
1994	Zoned Multiple Use	Natural Heritage	8243.5
1998	No Take	Natural Heritage	0.4
1968	Uniform Multiple Use	Sustainable Production	13.8
1974	Uniform Multiple Use	Natural Heritage	1.9
2007	Uniform Multiple Use	Natural Heritage	2.4
2001	No Access	Natural Heritage	2051.7
2006	Zoned w/No Take Areas	Natural Heritage	363686.7
1941	Uniform Multiple Use	Natural Heritage	25.8
1937	Uniform Multiple Use	Natural Heritage	18.8
1903	Uniform Multiple Use	Natural Heritage	24.2
2007	Uniform Multiple Use	Natural Heritage	22.9
2007	No Take	Natural Heritage	27.1
1908	No Access	Natural Heritage	1.9
1951	No Access	Natural Heritage	1.6
1972	Zoned w/No Take Areas	Natural Heritage	11.5
2007	Uniform Multiple Use	Natural Heritage	31.6
2007	No Take	Natural Heritage	17.3
1974	Uniform Multiple Use	Natural Heritage	2.8
2007	Uniform Multiple Use	Natural Heritage	22.0
1973	No Take	Natural Heritage	14.0
1974	Uniform Multiple Use	Natural Heritage	4.2
1962	Uniform Multiple Use	Natural Heritage	53.4
2007	Uniform Multiple Use	Natural Heritage	27.5
2007	No Take	Natural Heritage	25.3
1973	Uniform Multiple Use	Natural Heritage	0.0

Site Name	State	Government	Management Agency
Portugese Ledge State Marine Conservation Area	California	State	California Department of Fish and Game
Prime Hook National Wildlife Refuge	Delaware	Federal	U.S. Fish and Wildlife Service
Protection Island National Wildlife Refuge	Washington	Federal	U.S. Fish and Wildlife Service
Pupukea Marine Life Conservation District	Hawaii	State	Hawaii Department of Land and Natural Resources
Rachel Carson National Wildlife Refuge	Maine	Federal	U.S. Fish and Wildlife Service
Redwoods National Park ASBS State Water Quality Protection Area	California	State	California State Water Resources Control Board
Robert E. Badham ASBS State Water Quality Protection Area	California	State	California State Water Resources Control Board
Rookery Bay National Estuarine Research Reserve	Florida	Partnership	Florida Department of Environmental Protection
Rose Atoll National Wildlife Refuge	Pacific Islands	Federal	U.S. Fish and Wildlife Service
Sabine National Wildlife Refuge	Louisiana	Federal	U.S. Fish and Wildlife Service
Sachuest Point National Wildlife Refuge	Rhode Island	Federal	U.S. Fish and Wildlife Service
Salmon Creek Coast ASBS State Water Quality Protection Area	California	State	California State Water Resources Control Board
San Bernard National Wildlife Refuge	Texas	Federal	U.S. Fish and Wildlife Service
San Clemente Island ASBS State Water Quality Protection Area	California	State	California State Water Resources Control Board
San Diego-Scripps ASBS State Water Quality Protection Area	California	State	California State Water Resources Control Board
San Juan Channel and Upright Channel Special Management Fishery Area	Washington	State	Washington Department of Fish and Wildlife
San Miguel, Santa Rosa, and Santa Cruz Islands ASBS State Water Quality Protection Area	California	State	California State Water Resources Control Board
San Nicolas Island and Begg Rock ASBS State Water Quality Protection Area	California	State	California State Water Resources Control Board
San Pablo Bay National Wildlife Refuge	California	Federal	U.S. Fish and Wildlife Service
Santa Barbara and Anacapa Islands ASBS State Water Quality Protection Area	California	State	California State Water Resources Control Board
Saunders Reef ASBS State Water Quality Protection Area	California	State	California State Water Resources Control Board
Savage Neck Dunes Natural Area Preserve	Virginia	State	Virginia Department of Conservation and Recreation
Seatuck National Wildlife Refuge	New York	Federal	U.S. Fish and Wildlife Service
Shaw Island San Juan Islands Marine Preserve	Washington	State	Washington Department of Fish and Wildlife
Shell Keys National Wildlife Refuge	Louisiana	Federal	U.S. Fish and Wildlife Service
Siletz Bay National Wildlife Refuge	Oregon	Federal	U.S. Fish and Wildlife Service
Soquel Canyon State Marine Conservation Area	California	State	California Department of Fish and Game

Year Established	Level of Protection	Primary Conservation Focus	Site Area (km²)
2007	Uniform Multiple Use	Natural Heritage	27.6
1963	Uniform Multiple Use	Natural Heritage	39.6
1982	No Access	Natural Heritage	1.4
1983	Zoned w/No Take Areas	Natural Heritage	0.7
1966	Uniform Multiple Use	Natural Heritage	35.6
1974	Uniform Multiple Use	Natural Heritage	253.7
1974	Uniform Multiple Use	Natural Heritage	0.9
1978	Uniform Multiple Use	Natural Heritage	378.5
1973	No Access	Natural Heritage	158.5
1937	Uniform Multiple Use	Natural Heritage	581.2
1970	Uniform Multiple Use	Natural Heritage	1.0
1974	Uniform Multiple Use	Natural Heritage	5.9
1968	Zoned w/No Take Areas	Natural Heritage	14.6
1974	Uniform Multiple Use	Natural Heritage	199.5
1974	Uniform Multiple Use	Natural Heritage	0.4
1972	Uniform Multiple Use	Sustainable Production Restrictions	40.5
1974	Uniform Multiple Use	Natural Heritage	1113.5
1974	Uniform Multiple Use	Natural Heritage	258.3
1974	Uniform Multiple Use	Natural Heritage	36.9
1974	Uniform Multiple Use	Natural Heritage	141.4
1974	Uniform Multiple Use	Natural Heritage	3.0
1999	Zoned Multiple Use	Natural Heritage	0.0
1968	No Access	Natural Heritage	0.9
1990	Uniform Multiple Use	Natural Heritage	1.8
1907	Uniform Multiple Use	Natural Heritage	0.0
1991	No Access	Natural Heritage	4.3
2007	Uniform Multiple Use	Natural Heritage	59.6

Site Name	State	Government	Management Agency
South Puget Sound Wildlife Area	Washington	State	Washington Department of Fish and Wildlife
Southeast Santa Catalina Island ASBS State Water Quality Protection Area	California	State	California State Water Resources Control Board
St. Marks National Wildlife Refuge	Florida	Federal	U.S. Fish and Wildlife Service
St. Vincent National Wildlife Refuge	Florida	Federal	U.S. Fish and Wildlife Service
Stewart B. McKinney National Wildlife Refuge	Connecticut	Federal	U.S. Fish and Wildlife Service
Sund Rock Conservation Area	Washington	State	Washington Department of Fish and Wildlife
Supawna Meadows National Wildlife Refuge	New Jersey	Federal	U.S. Fish and Wildlife Service
Susquehanna National Wildlife Refuge	Maryland	Federal	U.S. Fish and Wildlife Service
Swanquarter National Wildlife Refuge	North Carolina	Federal	U.S. Fish and Wildlife Service
Sweetwater Marsh National Wildlife Refuge	California	Federal	U.S. Fish and Wildlife Service
Target Rock National Wildlife Refuge	New York	Federal	U.S. Fish and Wildlife Service
Ten Thousand Islands National Wildlife Refuge	Florida	Federal	U.S. Fish and Wildlife Service
Thunder Bay National Marine Sanctuary and Underwater Preserve	Michigan	Federal	National Marine Sanctuaries
Trinidad Head ASBS State Water Quality Protection Area	California	State	California State Water Resources Control Board
U-1105 Black Panther Historic Shipwreck Preserve	Maryland	Partnership	Navy/St. Mary's County Department of Recreation and Parks
Vandenberg State Marine Reserve	California	State	California Department of Fish and Game
Virgin Islands Coral Reef National Monument	U.S. Virgin Islands	Federal	National Park Service
Virgin Islands National Park	U.S. Virgin Islands	Federal	National Park Service
Waccamaw National Wildlife Refuge	South Carolina	Federal	U.S. Fish and Wildlife Service
Wallops Island National Wildlife Refuge	Virginia	Federal	U.S. Fish and Wildlife Service
Waquoit Bay National Estuarine Research Reserve	Massachusetts	Partnership	Massachusetts Department of Conservation and Recreation
Wertheim National Wildlife Refuge	New York	Federal	U.S. Fish and Wildlife Service
West Hawaii Regional Fishery Management Area	Hawaii	State	Hawaii Department of Land and Natural Resources
Western Santa Catalina Island ASBS State Water Quality Protection Area	California	State	California State Water Resources Control Board
White Rock (Cambria) State Marine Conservation Area	California	State	California Department of Fish and Game
Willapa National Wildlife Refuge	Washington	Federal	U.S. Fish and Wildlife Service
Yellow and Low Islands San Juan Islands Marine Preserve	Washington	State	Washington Department of Fish and Wildlife
Yukon Delta National Wildlife Refuge	Alaska	Federal	U.S. Fish and Wildlife Service
Zella M. Schultz/Protection Island Seabird Sanctuary	Washington	Partnership	Washington Department of Fish and Wildlife

Year Established	Level of Protection	Primary Conservation Focus	Site Area (km²)
1988	No Access	Natural Heritage	0.3
1974	Uniform Multiple Use	Natural Heritage	11.2
1931	Uniform Multiple Use	Natural Heritage	308.8
1968	Uniform Multiple Use	Natural Heritage	49.4
1985	Uniform Multiple Use	Natural Heritage	4.5
1994	No Take	Natural Heritage	0.3
1934	Uniform Multiple Use	Natural Heritage	17.9
1939	Uniform Multiple Use	Natural Heritage	0.0
1932	Uniform Multiple Use	Natural Heritage	67.1
1988	No Take	Natural Heritage	0.0
1967	Uniform Multiple Use	Natural Heritage	0.3
1996	Uniform Multiple Use	Natural Heritage	141.5
2000	Zoned Multiple Use	Cultural Heritage	1160.0
1974	Uniform Multiple Use	Natural Heritage	1.2
1993	Uniform Multiple Use	Cultural Heritage	0.1
1994	No Take	Natural Heritage	85.3
2001	Zoned Multiple Use	Natural Heritage	51.8
1956	Zoned w/No Take Areas	Natural Heritage	22.9
1997	Uniform Multiple Use	Natural Heritage	213.4
1971	Uniform Multiple Use	Natural Heritage	26.0
1988	Uniform Multiple Use	Natural Heritage	11.5
1947	Uniform Multiple Use	Natural Heritage	11.6
1999	Zoned Multiple Use	Sustainable Production	160.5
1974	Uniform Multiple Use	Natural Heritage	9.1
2007	Uniform Multiple Use	Natural Heritage	7.7
1936	Uniform Multiple Use	Sustainable Production	19.8
1990	Uniform Multiple Use	Natural Heritage	0.8
1980	Uniform Multiple Use	Natural Heritage	11688.1
1975	No Access	Natural Heritage	0.0

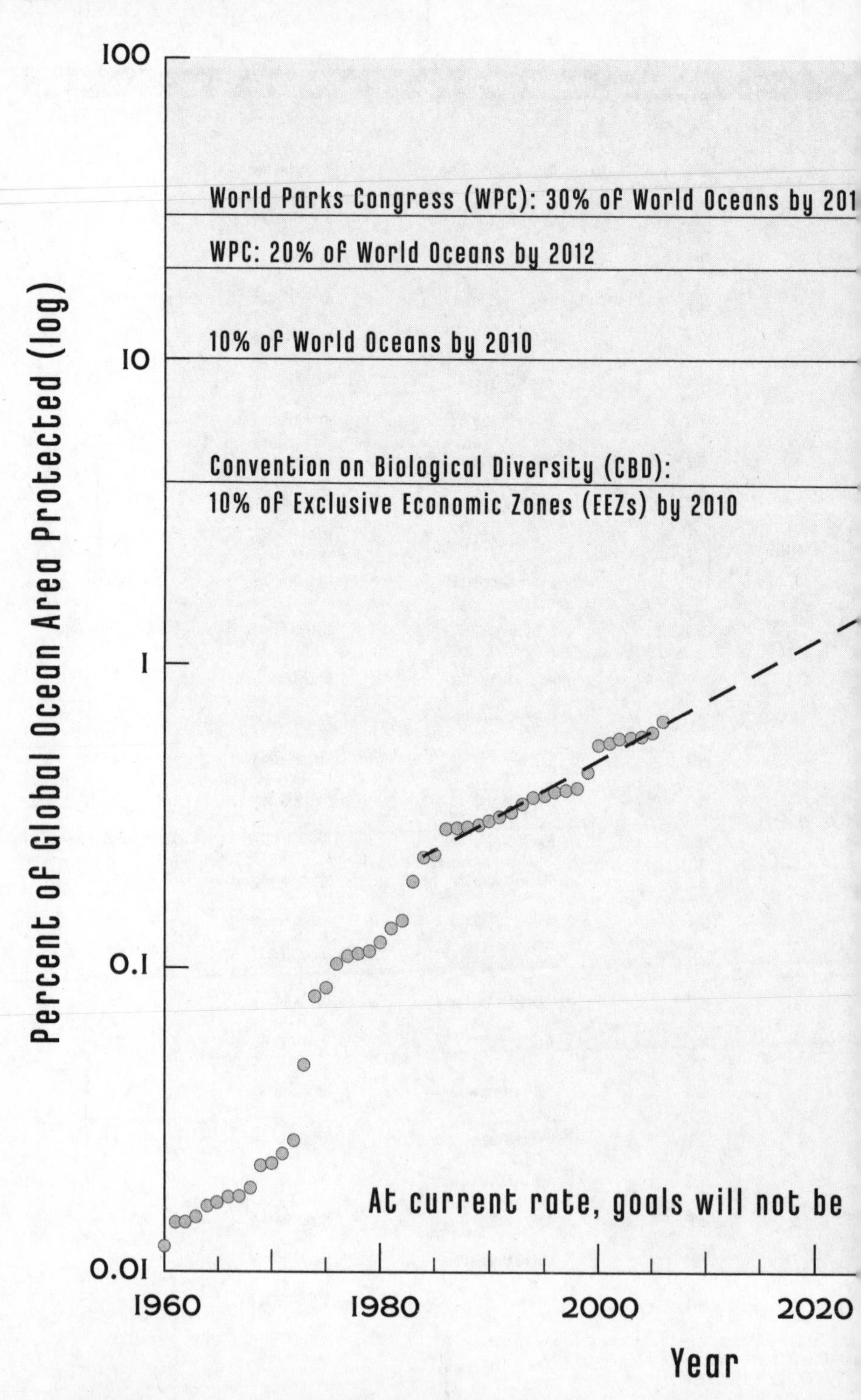
EXPECTED TARGET ATTAINMENT DATES
BASED ON CURRENT GROWTH
100
10
1
0.1
0.01
Percent of Global Ocean Area Protected (log)
World Parks Congress (WPC): 30% of World Oceans by 201
WPC: 20% of World Oceans by 2012
10% of World Oceans by 2010
Convention on Biological Diversity (CBD):
10% of Exclusive Economic Zones (EEZs) by 2010
At current rate, goals will not be
1960
1980
2000
2020
Year

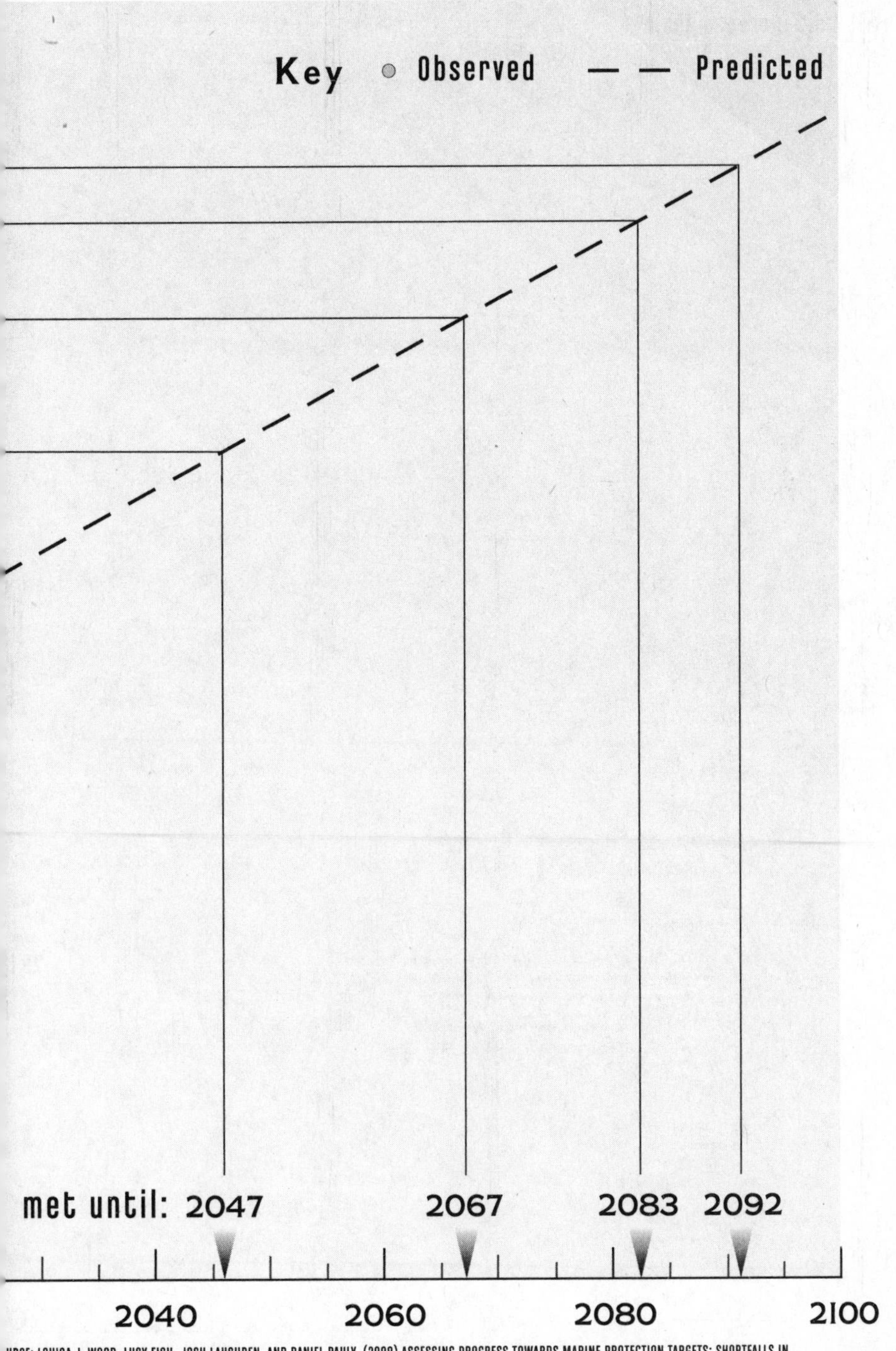

URCE: LOUISA J. WOOD, LUCY FISH, JOSH LAUGHREN, AND DANIEL PAULY. (2008) ASSESSING PROGRESS TOWARDS MARINE PROTECTION TARGETS: SHORTFALLS IN
FORMATION AND ACTION. ORYX, 42(3): 340-351.

MARINE PROTECTED AREAS (MPAs) WITH FULL PROTECTION BY MID-2008

2008: 450 MPAs, 300,000 square kilometers, 0.08% of World Oceans, 0.2% of Exclusive Economic Zones (EEZs)

(2009: No Increase yet)

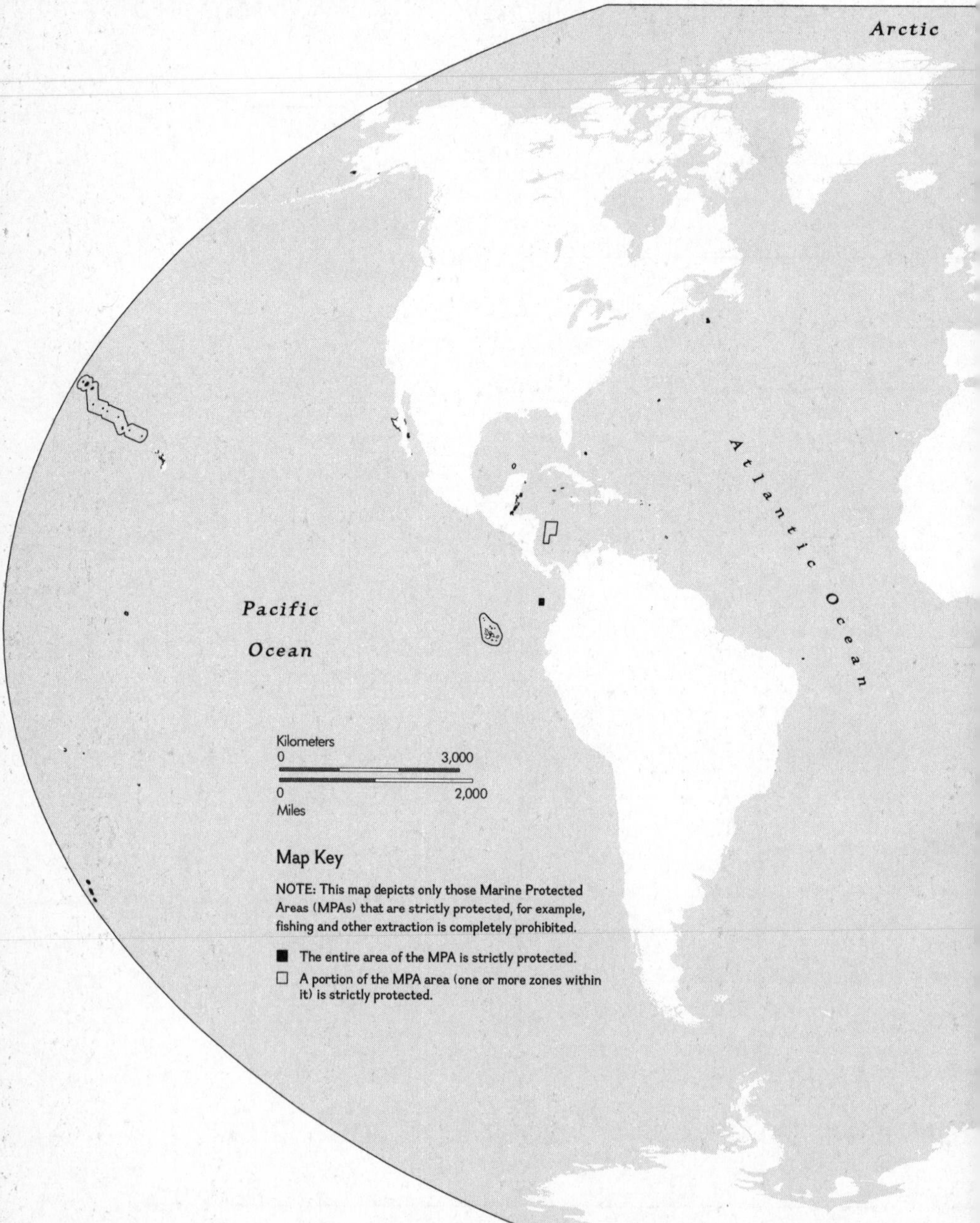

Arctic

Atlantic Ocean

Pacific Ocean

Kilometers
0 3,000
0 2,000
Miles

Map Key

NOTE: This map depicts only those Marine Protected Areas (MPAs) that are strictly protected, for example, fishing and other extraction is completely prohibited.

■ The entire area of the MPA is strictly protected.

□ A portion of the MPA area (one or more zones within it) is strictly protected.

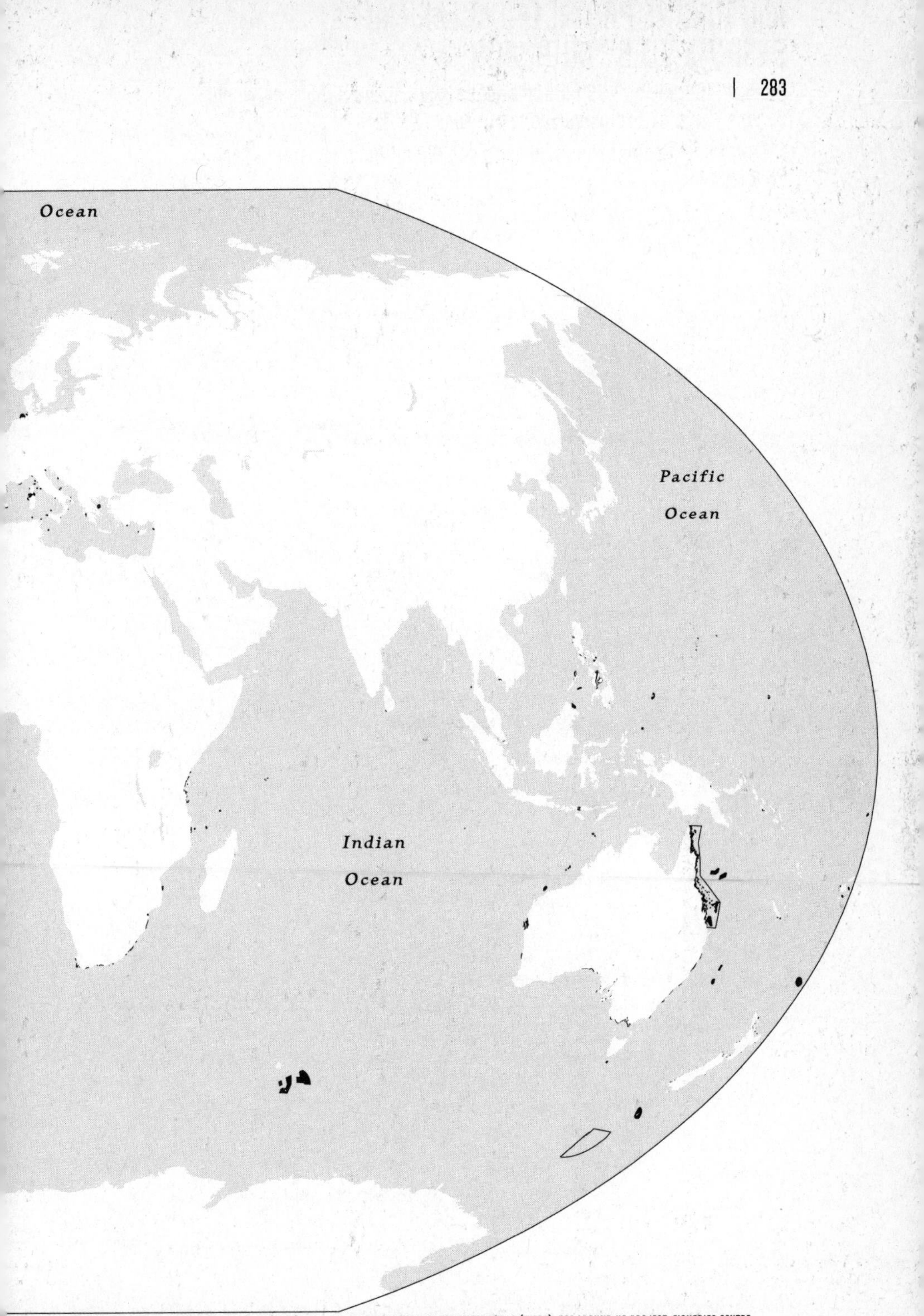

URCE: INTERNATIONAL UNION FOR CONSERVATION OF NATURE (IUCN); IUCN WORLD COMMISSION ON PROTECTED AREAS (WCPA); SEA AROUND US PROJECT, FISHERIES CENTRE,
VERSITY OF BRITISH COLUMBIA; UNITED NATIONS ENVIRONMENT PROGRAMME (UNEP); UNEP WORLD CONSERVATION MONITORING CENTRE (WCMC);
RLD WILDLIFE FUND (WWF)

ALL MARINE PROTECTED AREAS (MPAs) DESIGNATED BY MID-2008

2008: 4,400 MPAs, 2.35 million square kilometers, 0.65% of World Oceans, 1.6% of Exclusive Economic Zones (EEZs)

(2009: Bush Monuments Increased the Global Ocean Area Protected to 0.85%)

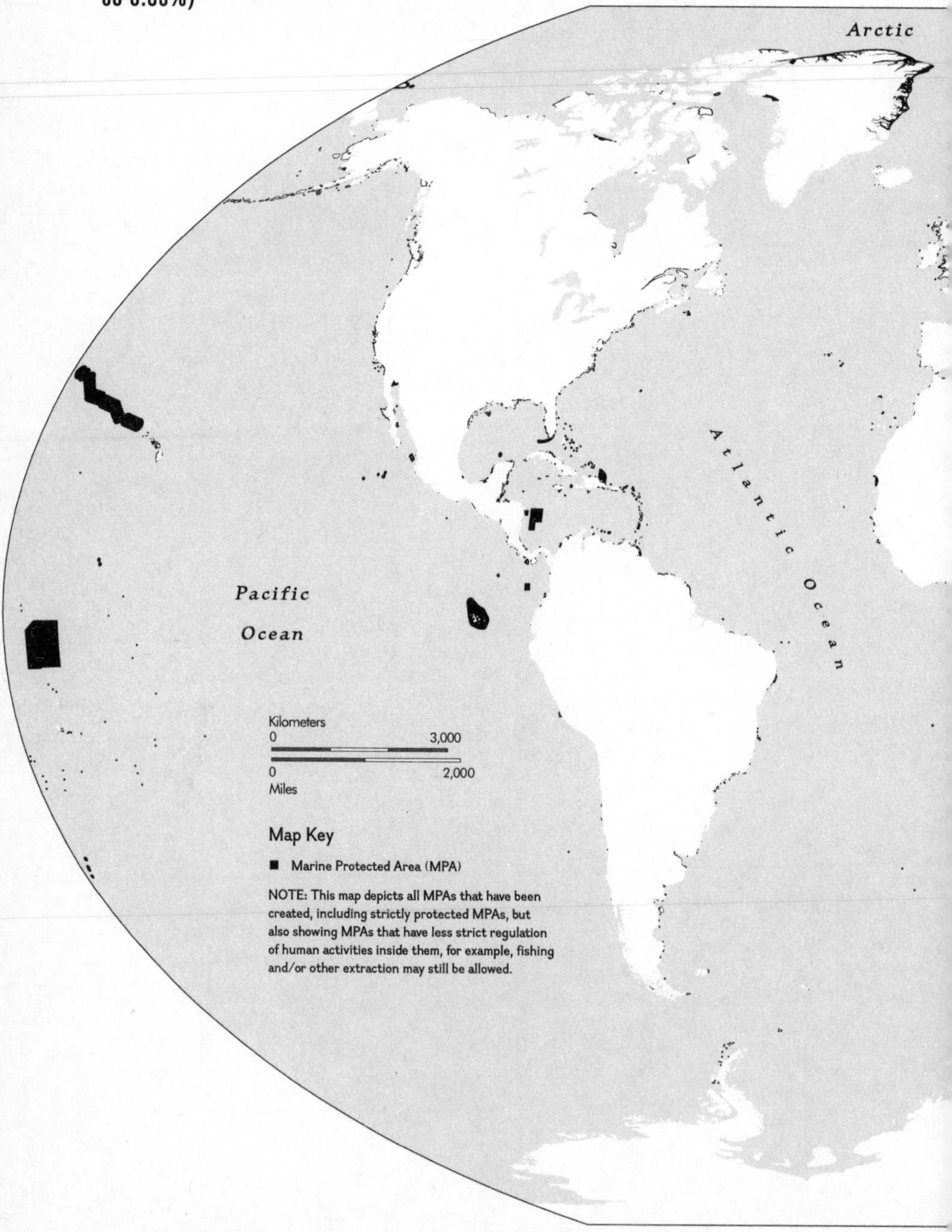

Arctic

Atlantic Ocean

Pacific Ocean

Kilometers
0 3,000
0 2,000
Miles

Map Key

■ Marine Protected Area (MPA)

NOTE: This map depicts all MPAs that have been created, including strictly protected MPAs, but also showing MPAs that have less strict regulation of human activities inside them, for example, fishing and/or other extraction may still be allowed.

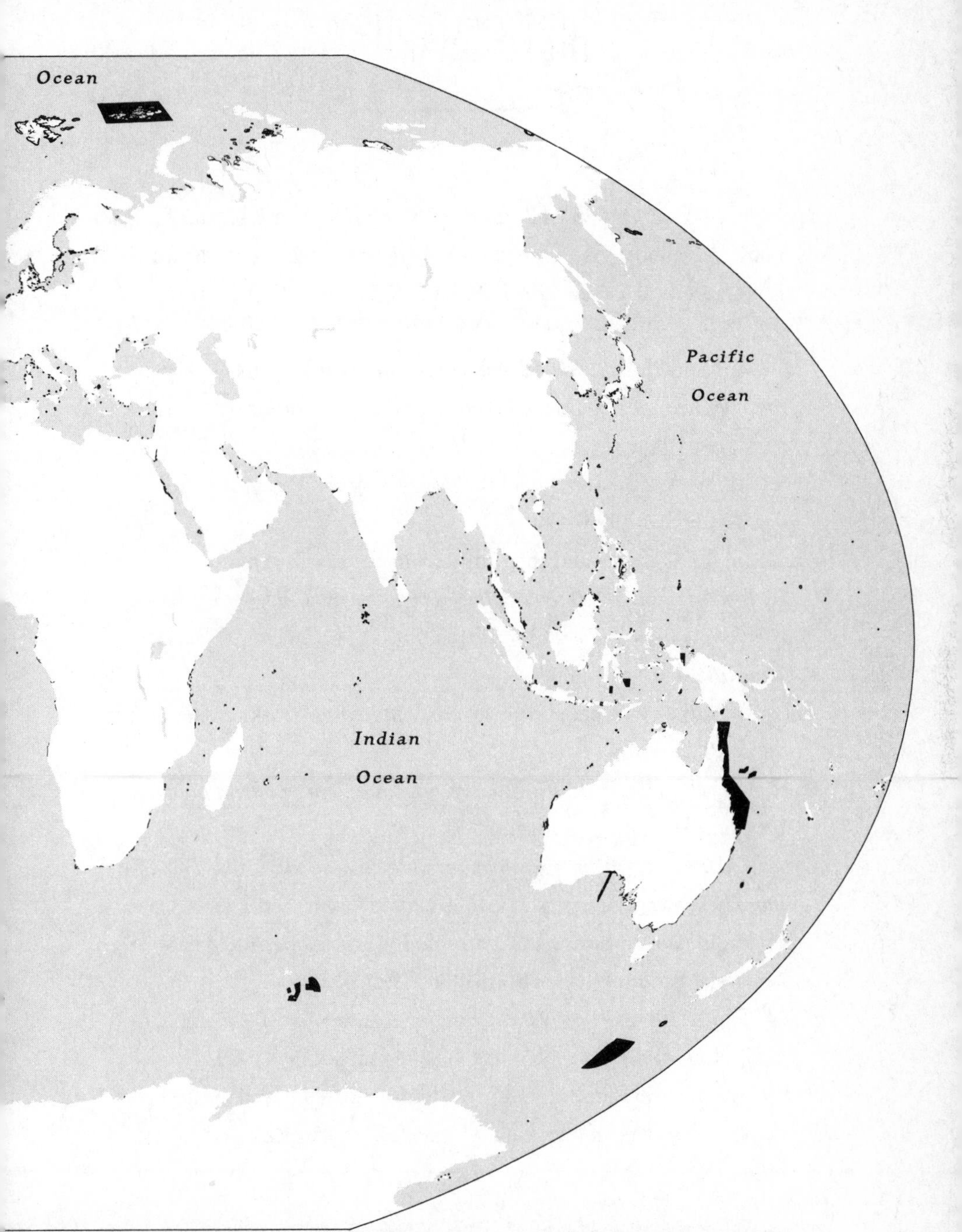

[SOU]RCE: INTERNATIONAL UNION FOR CONSERVATION OF NATURE (IUCN); IUCN WORLD COMMISSION ON PROTECTED AREAS (WCPA); SEA AROUND US PROJECT, FISHERIES CENTRE, [UNI]VERSITY OF BRITISH COLUMBIA; UNITED NATIONS ENVIRONMENT PROGRAMME (UNEP); UNEP WORLD CONSERVATION MONITORING CENTRE (WCMC); [AN]D WORLD WILDLIFE FUND (WWF)

BIBLIOGRAPHY

Alverson, Dayton L., Mark H. Freeberg, Steven A. Murawski, and J. G. Pope. "A Global Assessment of Fisheries Bycatch and Discards." FAO Technical Paper 339, Food and Agriculture Organization of the United Nations, 1994.

Andrews, Kenneth R. *Trade, Plunder and Settlement: Maritime Enterprise and the Genesis of the British Empire, 1480-1630.* Cambridge University Press, 1984.

Angel, Martin V. "Biodiversity of the Pelagic Ocean." *Conservation Biology* (December 1993), 760-772.

Anrady, A. L. "Plastics and their impacts in the marine environment." In *Proceedings of the International Marine Debris Conference on Derelict fishing Gear and the Ocean Environment, August 6-11, Honolulu, Hawaii,* 2000.

Anrandy, Anthony L., ed. *Plastics and the Environment.* John Wiley and Sons, 2003.

Baird, Rachel J. *Aspects of Illegal, Unreported and Unregulated Fishing in the Southern Ocean.* Springer, 2006.

Bakun, A. *Patterns in the Ocean: Ocean Processes and Marine Population Dynamics.* University of California Sea Grant, San Diego, California, in cooperation with Centro de Investigaciones Biológicas de Noroeste, La Paz, Baja California Sur, Mexico, 1996.

Balog, James. *Extreme Ice Now: Vanishing Glaciers and Changing Climate: A Progress Report.* National Geographic Society, 2009.

Baum, Julia K., and Ransom A. Myers. "Shifting baseline and the decline of pelagic sharks in the Gulf of Mexico." *Ecology Letters* (2004), 135-145.

Beebe, William. *Half Mile Down.* Harcourt, Brace and Company, 1934.

Bigg, Grant R. *The Oceans and Climate.* Cambridge University Press, 1996.

Blankenship, K. "Large Sanctuaries Urged for Recovery of Wild Oyster Population." *Chesapeake Bay Journal* (April 3, 2009). Available online at *http://www.bayjournal.com/article.cfm?article=3561*

Blatt, Harvey. *America's Food: What You Don't Know About What You Eat.* MIT Press, 2008.

Block, Barbara A., and E. Donald Stevens, eds. *Tuna: Physiology, Ecology, and Evolution.* Academic Press, 2001.

Block, Barbara A., et al. "Electronic Tagging and Population Structure of Atlantic Bluefin Tuna." *Nature* (April 28, 2005), 1121-1127.

Boersma, P. Dee, and Julia K. Parrish. "Limiting Abuse: Marine Protected Areas, a Limited Solution." *Ecological Economics* (November 1999), 287-304.

Borgese, Elisabeth Mann. *Seafarm: the Story of Aquaculture.* Harry N. Abrams, 1980.

Botsford, Louis W., Juan Carlos Castilla, and Charles H. Peterson. "The Management of Fisheries and Marine Ecosystems. *Science* (July 25, 1997), 509 – 515.

Brander, K. M. "Global Fish Production and Climate Change" (Climate Change and Food Security Special Feature). *Proceedings of the National Academy of Sciences* (December 11, 2007), 19709-19714.

Bromley, Daniel W. *Environment and Economy: Property Rights and Public Policy.* Basil Blackwell, 1991.

Brown, Lester R. *Plan B 3.0: Mobilizing to Save Civilization.* W. W. Norton and Company, 2008.

Burroughs, William, ed. *Climate: Into the 21st Century.* Cambridge University Press, 2003.

Carey, F. G. "Fishes with warm bodies." *Scientific American* (February 1973), 36-44.

Carr, Archie. *The Windward Road: Adventures of a Naturalist on Remote Caribbean Shores*. Alfred A. Knopf, 1956.

Carson, Rachel. *Lost Woods: The Discovered Writing of Rachel Carson.* Beacon Press, 1998.

———. *The Sea Around Us.* Oxford University Press, 1951.

Clark, Colin W. "The Economics of Overexploitation." *Science* (August 17, 1973), 630-634.

Clover, Charles. *The End of the Line: How Overfishing Is Changing the World and What We Eat*. New Press, 2006.

Colborn, Theo, Dianne Dumanoski, and John Peter Meyers. *Our Stolen Future: Are We Threatening Our Fertility, Intelligence, and Survival?—A Scientific Detective Story.* Plume, 1997.

Convention on Biological Diversity. *Decisions Adopted by the Conference of the Parties to the Convention on Biological Diversity at Its Ninth Meeting (Decision XI/20, Annexes I-III).* Convention on Biological Diversity, 2008.

Costello, Christopher, Steven D. Gaines, and John Lynham. "Can Catch Shares Prevent Fisheries Collapse?" *Science* (September 19, 2008), 1678-1681.

Cunningham, J. T. *The Natural History of the Marketable Marine Fishes of the British Islands.* Macmillan and Company, 1896.

Cushing, D. H. *The Provident Sea*. Cambridge University Press, 1988.

Daskalov, Georgi M., et al. "Trophic Cascades Triggered by Overfishing Reveal Possible Mechanisms of Ecosystem Regime Shifts." *Proceedings of the National Academy of Sciences* (June 19, 2007), 10518-10523.

Diamond, Jared. *Collapse: How Societies Choose to Fail or Succeed.* Penguin Books, 2005.

Diamond, Sandra L. "Bycatch Quotas in the Gulf of Mexico Shrimp Trawl Fishery: Can They Work?" *Reviews in Fish Biology and Fisheries* (June 2004), 207-237.

Dukes, J. "Burning Buried Sunshine: Human Consumption of Ancient Solar Energy," *Climatic Change* (2003), 31-44.

Earle, Sylvia A. *Lessons from History's Biggest Oil Spill.* Cosmos Club, 1992.

———. *Sea Change: A Message of the Oceans.* G.P. Putnam's Sons, 1995.

———. "The Search for Sustainable Seas." In *Fish, Aquaculture and Food Security: Sustaining Fish as a Food Supply*, ed. A. G. Brown. Record of a conference conducted by the ATSE Crawford Fund, Parliament House, Canberra, 2004, 13-19.

Earle, Sylvia A., and Al Giddings. *Exploring the Deep Frontier: The Adventure of Man in the Sea.* National Geographic Society, 1980.

Earle, Sylvia A., and Linda K. Glover. *Ocean, An Illustrated Atlas.* National Geographic Society, 2009.

Earle, Sylvia A., and Wolcott Henry. *Wild Ocean: America's Parks Under the Sea.* National Geographic Society, 1999.

Ellis, Richard. *Tuna: A Love Story.* Alfred A. Knopf, 2008.

Ellis, Richard, and John E. McCosker. *Great White Shark.* Stanford University Press, 1995.

Ernst, Howard R. *Chesapeake Bay Blues: Science, Politics, and the Struggle to Save the Bay*. Rowman and Littlefield, 2003.

Estes, James A., et al., eds. *Whales, Whaling, and Ocean Ecosystems.* University of California Press, 2007.

Fackler, Martin. "Waiter, There's Deer in My Sushi." *New York Times,* June 25, 2007.

Flannery, Tim. *The Weather Makers: How Man Is Changing the Climate and What It Means for Life on Earth.* Atlantic Monthly Press, 2006.

Food and Agriculture Organization of the United Nations Fisheries Department. *The State of the World Fisheries and Aquaculture.* Rome. FAO United Nations, 1995.

Francis, Daniel. *The Great Chase: A History of World Whaling.* Penguin Books, 1991.

Franklin, H. Bruce. *The Most Important Fish in the Sea.* Island Press, 2007.

Friedman, Thomas L. *Hot, Flat, and Crowded: Why We Need a Green Revolution and How It Can Renew America.* Farrar, Straus, and Giroux, 2008.

Game, Edward T., et al. "Pelagic Protected Areas: The Missing Dimension in Ocean Conservation." *Trends in Ecology and Evolution* (July 2009), 360-369.

Garstang, W. 1900. "The Impoverishment of the Sea." *Journal of the Marine Biological Association of the United Kingdom* (June 27, 2005), 1-69.

Glover, Linda K., and Sylvia A. Earle, eds. *Defying Oceans End: An Agenda for Action.* Island Press, 2004.

Grescoe, Taras. *Bottomfeeder: How to Eat Ethically in a World of Vanishing Seafood.* Bloomsbury, 2008.

Guinotte, John, and Victoria J. Fabry. "The Threat of Acidification to Ocean Ecosystems." *Current: The Journal of Marine Education* (2009), 2-7.

Halpern, Benjamin S., et al. "A Global Map of Human Impacts on Marine Ecosystems." *Science* (February 15, 2008), 948-952.

Hannesson, R. *Bioeconomic Analysis of Fisheries: An FAO Fishing Manual.* Wiley-Blackwell, 1993.

Hardin, Garrett. "The Tragedy of the Commons." *Science* (1968), 1243-1248.

Hardin, Garrett, ed. *Managing the Commons.* W. H. Freeman and Company, 1977.

Hassol, Susan Joy. *Impacts of a Warming Arctic—Arctic Climate Impact Assessment.* Cambridge University Press, 2004.

Hays, Graeme C., Anthony J. Richardson, and Carol Robinson. "Climate Change and Marine Plankton." *Trends in Ecology and Evolution* (June 1, 2005), 337-344.

Helvarg, David. *Blue Frontier: Dispatches from America's Ocean Wilderness.* Sierra Club Books, 2006.

———. *Blue Frontier: Saving America's Living Seas.* W. H. Freeman, 2001.

———. *50 Ways to Save the Ocean.* New World Library, 2006.

Herrick, Francis H. "The American Lobster: A Study of Its Habits and Development." *Bulletin of the U. S. Fisheries Commission* (1896), 1-252.

Heyerdahl, Thor. *The Ra Expeditions.* Doubleday, 1971.

Holt, Sidney J., and Lee M. Talbot. "New Principles for the Conservation of Wild Living Resources." *Wildlife Monographs* (April 1978) 3-33.

Hooker, Sascha K., and Leah R. Gerber. "Marine Reserves as a Tool for Ecosystem-Based Management: The Potential Importance of Megafauna." *BioScience* (January 2004), 27-39.

Hutchings, P. "Review of the Effects of Trawling on Macrobenthic Epifaunal Communities." *Australian Journal of Marine and Freshwater Research* (1990), 111-120.

Innis, Harold A. *The Cod Fisheries: The History of an International Economy.* University of Toronto Press, 1954.

International Union for Conservation of Nature. *Protected Areas of the World: A Review of National Systems, vol. 1- 4.* IUCN, 1991-1992.

International Union for Conservation of Nature-World Conservation Union, UN Environment Programme, and World Wide Fund for Nature. *Caring for the Earth: A Strategy for Sustainable Living.* IUCN, 1991.

Iudicello, Suzanne, Michael L. Weber, and Robert Wieland. *Fish, Markets, and Fishermen: The Economics of Overfishing.* Island Press, 1999.

Jackson, Jeremy B. C., et al. "Historical Overfishing and the Recent Collapse of Coastal Ecosystems." *Science* (July 27, 2001), 629-637.

Jones, Van. *The Green Collar Economy: How One Solution Can Solve Our Two Biggest Problems*. HarperOne, 2008.

Joseph, James, Witold Klawe, and Pat Murphy. *Tuna and Billfish: Fish Without a Country*. Inter-American Tropical Tuna Commission, 1998.

Kelleher, Graeme, ed. *Guidelines for Marine Protected Areas*. IUCN World Commission on Protected Areas, 1999.

Klingel, Gilbert. *The Bay*. The John Hopkins University Press, 1951.

Knecht, G. Bruce. *Hooked: Pirates, Poaching and the Perfect Fish*. Rodale, 2006.

Kurlansky, Mark. 2006. *The Big Oyster: History on the Half Shell*. Ballentine, 2006.

———. *Cod: A Biography of the Fish That Changed the World*. Alfred A. Knopf, 1997.

Kurzweil, Ray. *The Singularity Is Near: When Humans Transcend Biology*. Viking, 2005.

Larkin, P. A. "An Epitaph for the Concept of Maximum Sustainable Yield." *Transactions of the American Fisheries Society* (January 1977), 1-11.

Leopold, Aldo. *A Sand County Almanac: With Other Essays on Conservation From Round River*. Oxford University Press, 1966.

Lewison, R. L., et al. "Quantifying the Effects of Fisheries on Threatened Species: The Impact of Pelagic Longlines on Loggerhead and Leatherback Sea Turtles." *Ecology Letters* (March 2004), 221-231.

Lovelock, James. *Gaia: A New Look At Life on Earth*. Oxford University Press, 1979.

———. *The Revenge of Gaia*. Penguin Books, 2006.

Lubchenco, Jane, Stephen R. Palumbi, Steven D. Gaines, and Sandy Andelman. "Plugging a Hole in the Ocean: The Emerging Science of Marine Reserves." *Ecological Applications* (2003), S3-S7.

Lubchenco, J., et al. *The Science Marine Reserves*, 2nd ed. Partnership for Interdisciplinary Studies of Coastal Oceans, 2007.

Lynas, Mark. *Six Degrees: Our Future on a Hotter Planet.* National Geographic Society, 2008.

Macinko, Seth, and Daniel W. Bromley. *Who Owns America's Fisheries?* Island Press, 2002.

MacCracken, Michael C., et al., eds. *Prospects for Future Climate: A Special U.S./U.S.S.R. Report on Climate and Climate Change.* Lewis Publishers, 1990.

Mann, Charles C. *1491: New Revelations of the Americas Before Columbus.* Alfred A. Knopf, 2005.

Matteson, George. *Draggermen: Fishing on Georges Bank.* Four Winds Press, 1979.

Matthiessen, Peter. *Blue Meridian: The Search for the Great White Shark.* Random House, New York, 1971.

McDonald, Bernadette, and Douglas Jehl, eds. *Whose Water Is It? The Unquenchable Thirst of a Water-Hungry World.* National Geographic Society, 2003.

McGowan, J. A. "The Role of Oceans in Climate Change and the Ecosystem Effects of Change." In *Proceedings of the National Forum on Ocean Conservation,* 1991.

McIntosh, William Carmichael, and Arthur Thomas Masterman. *The Life-Histories of the British Marine Food-Fishes.* London, C. J. Clay and Sons, 1897.

McKibben, Bill. *The End of Nature.* Random House, 1989.

Melville, Herman. *Moby-Dick.* Richard Bentley, 1851.

Metz, Bert, et al., eds. *IPCC Special Report on Carbon Dioxide Capture and Storage: Summary for Policymakers and Technical Summary.* Cambridge University Press, 2005.

Miller, Kathleen A. "Climate Variability and Tropical Tuna: Management Challenges for Highly Migratory Fish Stocks." *Marine Policy* (January 2007), 56-70.

Moore, Charles J., Shelly L. Moore, Molly K. Leecaster, and Stephen B. Weisberg. "A Comparison of Plastic and Plankton in the North

Pacific Central Gyre." *Marine Pollution Bulletin* (December 2001), 1297-1300.

Murphy, Dallas. *To Follow the Water: Exploring the Ocean to Discover Climate*. Basic Books, 2007.

Myers, Ransom A., and Boris Worm. "Rapid Worldwide Depletion of Predatory Fish Communities." *Nature* (May 15, 2003), 280-283.

National Research Council. *An Assessment of Atlantic Bluefin Tuna*. National Academy Press, 1994.

———. *Conserving Biodiversity: A Research Agenda for Development Agencies*. National Academy Press, 1992.

———. *A Decade of International Climate Research: The First Ten Years of the World Climate Research Program*. National Academy Press, 1992.

———. *Oceanography in the Next Decade: Building New Partnerships*. National Academy Press, 1992.

———. *Sea-Level Change*. National Academy Press, 1990.

———. *Sustaining Marine Fisheries*. National Academy Press, 1999.

Norse, E. "Pelagic Protected Areas: The Greatest Park Challenge of the 21st Century." *Parks* (2006), 33-40.

Ocean Conservancy. *A Rising Tide of Ocean Debris and What We Can Do About It*. Ocean Conservancy, 2009.

Office of Technology Assessment. *Changing by Degrees: Steps to Reduce Greenhouse Gases*. U.S. Government Printing Office, 1991.

Ostrum, Elinor. *Governing the Commons: The Evolution of Institutions for Collective Action*. Cambridge University Press, 1990.

Ostrum, E. "The Rudiments of a Theory of the Origins, Survival, and Performance of Common Property Institutions." In *Making the Commons Work: Theory, Practice and Policy*, ed. Daniel W. Bromley. ICS Press, 1992.

Pacala, S., and R. Socolow. 2004. "Stabilization Wedges: Solving the Climate Problem for the Next 50 Years With Current Technologies." *Science* (August 13, 2004), 968-972.

Parsons, E. C. M., et al. "It's Not Just Poor Science—Japan's 'Scientific' Whaling May Be a Human Health Risk, Too." *Marine Pollution Bulletin* (September 2006), 1118-1120.

Pauly, Daniel, and Jay MacLean. *In a Perfect Ocean. The State of Fisheries and Ecosystems in the North Atlantic Ocean.* Island Press, 2003.

Pauly, Daniel, et al. "Towards Sustainability in World Fisheries." *Nature* (August 8, 2002), 689-695.

Reid, T. R. "The Great Tokyo Fish Market: Tsukiji." *National Geographic* (November 1995), 38-55.

Roberts, Callum M. "Effects of Fishing on the Ecosystem Structure of Coral Reefs." *Conservation Biology* (1995), 988-995.

———. *The Unnatural History of the Sea.* Island Press, 2007.

Roberts, Callum M., Julie P. Hawkins, and Fiona R. Gell. "The Role of Marine Reserves in Achieving Sustainable Fisheries." *Philosophical Transactions of the Royal Society: Biological Sciences* (2005), 123-132.

Rose, George A. "Cod spawning on a migration highway in the Northwest Atlantic." *Nature* (December 2, 1993), 458-461.

Rounsefell, George A., and W. Harry Everhart. *Fishery Science: Its Methods and Applications.* John Wiley and Sons, 1953.

Rozwadowski, Helen M. *Fathoming the Ocean: The Discovery and Exploration of the Deep Sea.* Harvard University Press, 2005.

Russell, E. S. *The Overfishing Problem.* Cambridge University Press, 1942.

Russell, Dick. *Striper Wars: An American Fish Story.* Island Press, 2005.

Safina, Carl. *Song for the Blue Ocean.* Henry Holt, 1997.

Sibert, John, John Hampton, Pierre Kleiber, and Mark Maunder. "Biomass, Size, and Trophic Status of Top Predators in the Pacific Ocean." *Science* (December 15, 2006), 1773-1776.

Simmons, Matthew R. *Twilight in the Desert: The Coming Saudi Oil Shock and the World Economy.* John Wiley and Sons, 2005.

Sobel, Jack, and Craig Dahlgren. *Marine Reserves: A Guide to Science, Design and Use.* Island Press, 2004.

Steneck, Robert S. "Human Influences on Coastal Ecosystems: Does Overfishing Create Trophic Cascades?" *Trends in Ecology and Evolution* (November 1, 1998), 429-430.

Steneck, Robert S., et al. "Kelp Forest Ecosystems: Biodiversity, Stability, Resilience and Future." *Environmental Conservation* (2002), 436-459.

Stolzenburg, William. *Where the Wild Things Were: Life, Death, and Ecological Wreckage in a Land of Vanishing Predators.* Bloomsbury Publishing, 2008.

Sumaila, Ussif Rashid, et al. "Potential Costs and Benefits of Marine Reserves in the High Seas." *Marine Biological Progress Series* (2007), 305-310.

Sutherland, W. J., et al. "One Hundred Questions of Importance to the Conservation of Global Biological Diversity." *Conservation Biology* (June 2009), 557-567.

Taylor, Harden F. *Survey of Marine Fisheries of North Carolina.* University of North Carolina Press, 1951.

Thompson, R. C., et al. "Lost at Sea. Where Is All the Plastic?" *Science* (May 7, 2004), 838.

Troubled Waters: A Special Report on the Sea. *The Economist* (December 30, 2008), 1-16.

United Nations World Commission on Environment and Development. *Our Common Future: Annex to General Assembly Document A/42/427.* United Nations, 1987.

Verity, Peter G., Victor Smetacek, and Theodore J. Smayda. "Status, Trends and the Future of the Marine Pelagic Ecosystem." *Environmental Conservation* (2002), 207-237.

Walford, Lionel A., ed. *Fishery Resources of the United States.* U.S. Fish and Wildlife Service. 1945.

Warner, William W. *Distant Water: The Fate of the North Atlantic Fisherman.* Little, Brown and Company, 1977.

Weisman, Alan. *The World Without Us.* St. Martins Press, 2007.

Wilson, Edward O. *The Creation: An Appeal to Save Life on Earth.* W. W. Norton and Company, 2006.

———. *The Diversity of Life.* Harvard University Press, 1992.

———. *The Future of Life.* Alfred A. Knopf, 2002.

Wood, Louisa J., Lucy Fish, Josh Laughren, and Daniel Pauly. "Assessing Progress Towards Global Marine Protection Targets: Shortfalls in Information and Action." *Oryx* (July 2008), 340-351.

Worm, Boris, et al. "Impacts of Biodiversity Loss on Ocean Ecosystem Services." *Science* (November 3, 2006), 787-790.

Worm, Boris, Heike K. Lotze, and Ransom A. Myers. "Predator Diversity Hotspots in the Blue Ocean." *Proceedings of the National Academy of* Sciences (August 19, 2003), 9884-9888.

Whynott, Douglas. *Giant Bluefin.* Farrar, Straus, and Giroux, 1995.

Yergin, D. *The Prize: The Epic Quest for Oil, Money, and Power.* Simon and Schuster, 1991.

Zaradic, Patricia, and Oliver R.W. Pergams. "Videophilia: Implications for Childhood Development and Conservation." *Journal of Developmental Processes* (Spring 2007), 130-144.

WEB LINKS

Aquarius: World's Only Undersea Research Center
NOAA's underwater laboratory where scientists can stay for up to ten days.
http://www.uncw.edu/aquarius/about/about.htm

Arctic Passage
Articles, photographs, and maps complementing *NOVA*'s program about the Franklin and Amundsen expeditions to the Northwest Passage and their dissimilar results.
http://www.pbs.org/wgbh/nova/arctic/

Beaufort Gyre Exploration Project
Scientists are studying the accumulation and release mechanism of fresh water and its role in climate variability.
http://www.whoi.edu/beaufortgyre/background.html

Blue Frontier
Chronicles the Sustainable Seas Expeditions, a joint project between the National Geographic Society and the National Oceanic and Atmospheric Administration (NOAA) to explore the United States' national marine sanctuaries.
http://www.nationalgeographic.com/seas/

Census of Marine Life
A global network of scientists from over 80 countries who are assessing and attempting to explain the diversity, distribution, and abundance of past, present, and future life in the oceans.
http://www.coml.org/

Center for Coastal & Ocean Mapping (CCOM) Joint Hydrographic Center
A national center for expertise in ocean mapping and hydrographic sciences, run by the University of New Hampshire.
http://www.ccom-jhc.unh.edu/

Coral Reef Alliance

An international organization working to promote knowledge and interest in coral reefs and their unique ecosystems to prevent further decline.
http://www.coral.org/

Coral Reef Conservation Program

NOAA's program to support science and management in order to preserve, restore, and sustain coral reefs.
http://coralreef.noaa.gov/

Deep Ocean Expeditions

Its mission is to educate the world about the deep sea through scientific research and adventure.
http://www.deepoceanexpeditions.com/index.html

Earth Observatory

Provides images and information about discoveries made by NASA scientists and includes satellite images, climate models, and in-the-field research.
http://earthobservatory.nasa.gov/

Environmental Defense Fund Seafood Selector

Helps consumers make smart choices when eating seafood.
http://www.edf.org/page.cfm?tagID=1521

Google Earth

Explore the ocean with Google Earth's latest feature.
http://earth.google.com/ocean/

IFREMER (French Research Institute for the Exploitation of the Sea)

The institute monitors the ocean, using a fleet of ships and underwater vessels, to contribute to knowledge about the ocean and its resources.
http://www.ifremer.fr/anglais/institut/missions.htm

International Arctic Buoy Programme

Information from a network of drifting buoys deployed in the Arctic Ocean to track meteorological and oceanographic information in real time.
http://iabp.apl.washington.edu/

International Union for Conservation of Nature

The oldest and largest global network of organizations trying to find solutions to environmental and development challenges.
http://www.iucn.org/about/

International Whaling Commission

An organization founded to ensure the conservation of world whale populations.
http://www.iwcoffice.org/

Marine Protected Areas of the United States

Provides detailed information about the marine protected areas, including maps.
http://mpa.gov/

McMurdo Station

McMurdo Station is the largest Antarctic station, set up in 1955 to be the hub of the U.S. Antarctic program.
http://www.nsf.gov/od/opp/support/mcmurdo.jsp

NASA/Goddard Space Flight Center Scientific Visualization Studio: Global Rotation Showing Seasonal Land Cover and Arctic Sea Ice

An animation of one full rotation of Earth, showing the variation in land cover and Arctic sea ice over time.
http://svs.gsfc.nasa.gov/vis/a000000/a003400/a003404/

NASA/Goddard Space Flight Center Scientific Visualization Studio: SeaWiFS Biosphere Data Over the North Pacific

Animations based on information collected from the Seastar satellite since 1997, including measurements of photosynthesis and ambient carbon.
http://svs.gsfc.nasa.gov/vis/a000000/a003400/a003471/

National Geographic: Animals

A database of information about different species from around the world.
http://animals.nationalgeographic.com/

National Marine Sanctuaries
Information about the nation's national marine sanctuaries from NOAA.
http://sanctuaries.noaa.gov/

National Snow and Ice Data Center
Part of the Cooperative Institute for Research in Environmental Sciences at the University of Colorado at Boulder, the center aims to collect and disseminate scientific data about the world's frozen realms.
http://nsidc.org/about/expertise/overview.html

Nature Conservancy: Coral Reefs
Information on coral reefs from an internationally recognized organization that promotes diversity of life on Earth.
http://www.nature.org/joinanddonate/rescuereef/

Ocean Explorer
NOAA's program to inspire interest in the ocean through new discovery.
http://www.oceanexplorer.noaa.gov/

Oceana
An international organization aimed at ocean conservation.
http://oceana.org/north-america/home/

Pacific Marine Environmental Laboratory
A part of NOAA carrying out scientific investigations of atmospheric and oceanographic processes.
http://www.pmel.noaa.gov/

Project GloBAL (Global Bycatch Assessment of Long-lived Species)
An organization that analyzes bycatch information for marine mammals, seabirds, and sea turtles.
http://bycatch.env.duke.edu/species/loggerhead

ReefIndia
A site run by the National Institute of Oceanography that showcases coral reef research being conducted in India.
http://reefindia.org/

SeamountsOnline

Information, collected since 2001, about the biology around seamounts.
http://pacific.sdsc.edu/seamounts/

Sustainable Seas Expeditions / Monterey Bay

The interactive website of Monterey Bay National Maritime Sanctuary.
http://www.nationalgeographic.com/monterey/ax/primary_fs.html

Tree of Life Web Project

A collaborative effort of enthusiasts and scientists from around the world to provide information about biodiversity, evolutionary history, and species characteristics of many different kinds of organisms.
http://www.tolweb.org/tree/

United Nations Atlas of the Oceans

An atlas of information on ocean issues for the public, for scientists, and for policymakers.
http://www.oceansatlas.org/index.jsp

USGS Water Resources of the United States

The U.S. Geological Survey website provides information about water, including water information by state, for the benefit of the nation's citizens.
http://water.usgs.gov/

USGS Water Science for Schools

A teaching tool for educators trying to provide information about water topics.
http://ga.water.usgs.gov/edu/

Vents Program

NOAA's program monitoring volcanoes and hydrothermal venting in the ocean.
http://www.pmel.noaa.gov/vents/index.html

Winds: Measuring Ocean Winds From Space

NASA's Jet Propulsion Laboratory uses radar to track wind and weather patterns in the world's oceans, as well as monitor glacial melting.
http://winds.jpl.nasa.gov/

Woods Hole Oceanographic Institution

The largest, private nonprofit ocean research, education, and engineering organization.
http://www.whoi.edu/

WoRMS (World Register of Marine Species)

The WoRMS database is working to provide a comprehensive and authoritative list of marine species.
http://www.marinespecies.org/about.php